TEACHING IN COMPREHENSIVE SCHOOLS

A SECOND REPORT

TEACHING IN
COMPREHENSIVE SCHOOLS

A SECOND REPORT

ISSUED BY THE
INCORPORATED ASSOCIATION OF
ASSISTANT MASTERS IN
SECONDARY SCHOOLS

CAMBRIDGE
AT THE UNIVERSITY PRESS
1967

Published by The Syndics of the Cambridge University Press
Bentley House, 200 Euston Road, London, N.W. 1
American Branch: 32 East 57th Street, New York, N.Y. 10022

Library of Congress Catalogue Card Number: 67–10987

Printed in Great Britain
at the University Printing House, Cambridge
(Brooke Crutchley, University Printer)

CONTENTS

FOREWORD

The members of the Assistant Masters Association, who serve in secondary schools of all kinds, maintained and independent, selective and non-selective, have a common purpose to secure for all children of secondary school age the educational opportunities which are best suited to their abilities, aptitudes and needs.

In 1960 the Association published its first report on 'Teaching in Comprehensive Schools' which set out to give a factual and objective description of the work then being done in comprehensive schools. It was based on the experience of members of the Association serving in such schools at that time. This second report, though based on information provided by a greater number of members, many of whom have now had much longer experience, is intended to serve the same purpose as the first report.

Although neither report attempts a comparative assessment of the advantages and disadvantages of secondary schools of different kinds, nor is intended as a statement of Association policy, the Association believes that the knowledge and experience embodied in these reports will be of value to all members of the Association, irrespective of the kind of secondary school in which they serve, and, indeed, to all who are interested in the development of secondary education.

The Executive Committee of the Association wishes to record its appreciation of the contribution made to this second report by all those who have completed the questionnaire and attended the conferences which provided the material on which this report is based. In particular the Executive Committee expresses its grateful thanks to

FOREWORD

Mr N. L. Clay, who has analysed a very large amount of information and shaped it into what we hope will prove to be a valuable and stimulating book.

A. W. S. HUTCHINGS
Secretary
Assistant Masters Association

1

OUR PURPOSE

The purpose of this second report is to present the views of members of the Association who teach in comprehensive schools. It is their book. Their figures, their opinions, give the report its value. What was said of the first report can be applied to its successor: 'It is presented as a factual and objective document, which does not attempt to assess the merits or de-merits of these schools.' Where any part seems to be a repetition of the first report, this is because the views of our members in 1965 are the same as those of our members in 1959. There has been no attempt to build on, to extend or to inflate the first report. The views of our members have been collected and presented as if there had been no first report.

This second report, like the first, is given as 'an interim report, since it is too early to make a full assessment of the work' of comprehensive schools.

It seeks 'to study afresh, in the light of wider and longer experience,' this approach to the organization of secondary education and sees in the number and nature of experiments very recently begun a need to wait a few years longer before there can be any full assessment. For example, the introduction of the Certificate of Secondary Education has only begun to make its impact. When the new pupils of September 1965 have taken a full course in a school organized to lead to the Certificate of Secondary Education, that is in the summer of 1970, then may be the time for a new survey.

It may help to state at once and in the plainest terms what this report does *not* attempt. It does *not* express, it does not try to express, the policy or official attitude of this Association. It would be very wrong to take as the mind of the Association any sentence quoted as the view of one of our members or any unquoted sentence supplied by the writer in his effort to give coherence and continuity to varied and fragmentary material. It is *not* an attack on comprehensive schools; it is *not* a defence of comprehensive schools. It is *not* a balance sheet or a Judge's summing up. It is *not* any part of 'the debate about the kind of society we want'. It is content to be a report of what experience of teaching in a comprehensive school has led our members to think and feel.

This book can be regarded as the work of more than 70 practising teachers, our members who have supplied valuable answers to a detailed questionnaire and who have attended regional conferences organized by this Association. There they supplemented their written contribution and spoke their mind, something different from answering questions. In a number of schools two or three members shared the work of writing answers. So at a conference a member who had written Section A often spoke on the material of Section B. We try here to pass on to our readers 'the feel of working inside a comprehensive school' which our members' written and spoken contributions gave us.

Issues in education lead to many opinions; decisions in education need to be based on informed opinions. In giving their experience, our members in comprehensive schools are making a practical contribution. There are aspects of teaching in a comprehensive school which only practising teachers know intimately. Opinion, if it is to be informed opinion, must pay respect to the theorists; equal respect at least is due to the practitioners. The figures our members supply give a factual basis to a practical contribution. Their views on how ideas work out in practice deserve serious consideration. We assemble and arrange the experience of our members to give readers a chance to look into the classrooms, to pass along the corridors, to come closer to boys and girls at work and play, the children for whom comprehensive schools were built. We can see these children in their schools not through the eyes of the Chief Education Officer as the Headmaster takes him round the school, nor through the eyes of the Headmaster, but through the eyes of the worker in the classroom and through the eyes of the leader whose time is divided between teaching and organizing. We have been fortunate in that much of our material comes from leaders, our members who occupy positions of authority and responsibility, those who both help to form decisions and translate those decisions into schemes of organization.

Because the report comes from practising teachers it should be of value to influential people who have a professional interest in education. In newspaper circles, both leader-writers and journalists specializing in educational matters will find here much to help them in their task of informing the public. An Inspector, local or national, who knows well the schools in his own area, can here learn of what goes on in other areas. In Colleges of Education and University Departments of Education, those who present a picture of state education today can here find practical illustrations. Administrators serving city or county can use material here to apply practical tests to plans before them. Our purpose, too, is to help the many amateurs who serve the community on Boards of Governors, Divisional Executives, and Education Committees. This report can help them to understand better the schools that exist now and the schools which are being planned.

Yet our first aim is to help teachers. All teachers of children of secondary school age should be interested in and stimulated by accounts by other teachers of important experiments now being tried in comprehensive schools. No teacher can afford to ignore what succeeds with children in comprehensive schools; any teacher should be wary of trying something which has not worked well in another kind of school. Teachers in comprehensive schools will appreciate the opportunity to compare their ways with the ways of other teachers in other areas. Teachers in modern schools will find much to help them with their pupils, both the more gifted and the less able 'Newsom' children. Grammar school teachers will note and profit from different ways of arranging work and courses for similar children in comprehensive schools. Any teacher with an open mind can find here something to help him to be a better teacher. For some time ahead, there will be in England different kinds of schools for children between 11 and 18, schools within the state framework and schools outside that framework. In the state sector, co-existence will continue. The need will remain for teachers to study what is done in school community and in classroom in other kinds of school so that cross-fertilization can take place.

The activity in response to Circular 10/65 from the Department of Education and Science asking for plans for reorganization of secondary schools will affect almost all engaged in secondary education. Even teachers in existing comprehensive schools may find that new plans propose to change the catchment area which supplies pupils. The number of plans accepted is likely to be smaller than the number of plans submitted, so that in the first stages there will be only a minority of grammar schools and modern schools not affected. Thus teachers need to know as much as possible about what serving in a comprehensive school means so that they can offer constructive criticism of plans involving their own schools. Opposition to change because it means disturbance is both unworthy of a profession and certain to be ineffectual.

Many teachers will be called upon to act as representatives of bodies of teachers, on local working parties and on committees within associations. This book will help such people to assess not only the plans submitted by Local Education Authorities but also the responses to the plans from other teachers.

'In any reorganization plans, the life of the teacher will almost certainly undergo profound changes, academic and personal. It is therefore particularly important that he should be allowed freedom to express his views upon the proposals.' There is no point in naming the group of associations from which these words came; any and every teachers' organization would subscribe to the view. Discussion and re-appraisal after considering what discussion brings up is an essential feature of the democratic way of life. The material here, operating on active and open minds, can lead to sounder discussion by reducing the element of emotionalism and prejudice (in the sense of 'having one's mind made up before considering the evidence') in discussion. The words of the Chairman of this Association (29 December 1965), speaking for himself, not as a spokesman for others, are to the point here:

Those who teach in comprehensive schools, and they alone, know precisely what reorganization means in terms of its effects upon the pupils and the teachers. They alone can assess its consequences. They alone (not political theorists, nor parents, nor administrators) can tell the nation what in simple, unvarnished truth reorganization really means. And it is to them, rather than to anyone else, that the public would do well to listen.

2

THE BACKGROUND

NOTIONS IN THE AIR

This current picture of life in comprehensive schools should be set against the background of notions in the air. One notion, growing in force, is that 'streaming' children into separate forms according to their ability does not do the good it aims at doing and may do harm in ways which have not been foreseen. Parents of pupils in good streams or top streams accept streaming; parents of pupils in the lowest streams seem to keep quiet. The debate is waged mainly among teachers. This report shows to what extent streaming is on trial in comprehensive schools. The teachers opposing streaming do so for reasons strikingly like those influencing opponents of separate grammar and modern schools—dislike of rigid separation based on early decisions. This dislike causes some to reorganize secondary education in the manner of the Leicestershire plan and others to try further variations in the two-tier system. 'A common school up to 16' cry some, with an eye to a sixth-form college recruiting from a number of common schools. Others see the comprehensive school as a common school for the full length of school life. What the different planners have in common is a desire (if not determination) to end the system of having in one area, side by side, schools of differing kinds for the children of 11 upwards. So long as you have a grammar school and modern schools dividing the children of 11–15 between them, one school will be equated in public opinion with 'superior' and 'success', the other with 'inferior' and 'failure'. The people in general want

6

grammar school children to have grammar school education. Later sections will suggest that there is now less doubt about whether a comprehensive school will supply this. Indeed, some people want the standards and benefits of grammar school education to be extended to all. They hope that through comprehensive schools this can be done.

The current feeling is against starting children of 11 in a separate school because someone somehow decided they were not good enough to go to a different separate school. The battle has ceased to be about the form the 11-plus should take. Few outsiders realize in how many areas a distinction is drawn without relying on single-day written examinations in Arithmetic and English, with or without intelligence tests. Traces remain, but only traces, of earlier faith in intelligence tests. Indeed, to quote our Chairman again: 'In plain, unvarnished fact, the margin of error in selection is now known to be so wide that the nation has lost much of its earlier faith in the whole process.'

The real testing point of selection at 11-plus for different kinds of schools is possibly that border line, the last child above which goes to a grammar school, the first one below to a modern school.

The public has come to see not only that the difference between them is extremely slight, but that there is no question of standard which a boy or girl must reach and which is applied or can be applied uniformly over England and Wales. There are only so many places in grammar schools; when they are filled, the rest of the age group go elsewhere. So in Area *A*, where there are places for 22 per cent, the first to be excluded may be of superior quality to the last to be included in Area *B*, which provides places for 26 per cent.

At this testing point, the quality of the school whose intake begins where that of the grammar school stops is of great importance. In some areas only exaggeration can make the best pupil who goes to the modern school a 'deprived' child because that school has good buildings, good teachers and a sense of purpose. Ten years ago the modern schools preparing a form for 'O' Levels in G.C.E. could be compared to a trickle; now they form a steady stream. Excellent results from the best modern schools have invited comparison with the results of the lowest form of the neighbouring grammar school. Some will hold that comparisons of this sort strengthen the case for bringing the bottom form of the one and the top form of the other into the fold of a comprehensive school. People in general are not always consistent; one can hear the same people praising modern schools for successes in external examinations (as they should) and insisting that a child who goes to a modern school is branded from the start as a failure, that he feels a failure and is not allowed to forget that he was not good enough for a grammar school. Add salt to taste.

The introduction of the C.S.E. is in the process of making its impact. Nineteen sixty-six will be the first year of widespread entries. Later pages refer to the pleasure felt in comprehensive schools after tentative entries. The rating of C.S.E. by employers has not yet been put to the test. It may be that we shall hear less of the claim that education in a modern school restricts the area of opportunity in employment. It may be that the lowest form of a grammar school will show improvement in effort and attitude. It seems now that the larger school can more comfortably provide mixed courses so that the individual boy of the border-line level can limit his 'O' Level entry to his best subjects and supplement that entry by taking other subjects in the C.S.E.

Ideas about the length of school life and the right point for making a change from primary school to secondary school are in the process of changing. Ahead of us is the raising of the school-leaving age, a tremendous step depending on accommodation, teachers, and money to provide both. Our members in comprehensive schools are confident that they already have experience of providing special courses for the boys of 15–16 and can extend the provision to meet the needs of larger numbers.

Many of them would prefer to wait to see the effects of the use of the C.S.E. examination. These effects may strengthen the case for relying on voluntary staying on after 15. When school life does go on until all the pupils are 16, it does not follow that the span of 11–16 will be the basis of organization. Some now advocate making the age of transfer 12 instead of 11. At the time of writing, the report of the Plowden Committee which is investigating this problem is awaited. Its conclusions and recommendations may make many Local Authorities reconsider plans for organizing secondary education based on divisions such as 5–10, 11–16.

The 'O' Level itself is coming under the fire of snipers. If some thinkers get their way it will cease to be the terminal examination for an important proportion of the school population. In a country not outstanding for speed in making radical changes it will take time to win sufficiently strong support for only two testing points, C.S.E. at sixteen (or earlier) and 'A' Levels at eighteen. So long as there is a possibility that this may be the pattern of the future, it must be taken into account when reorganization is being considered.

WHAT PARENTS THINK

Discussion of the attitude of parents to comprehensive schools brings us out of the upper air of theory. Our members in these schools have their feet on the ground and can speak from experience on this matter. The fears of pessimists have been proved to be much exaggerated. As parents see the comprehensive schools deliver the goods, that is, see their children succeed in external examinations and, helped by this success, proceed to colleges or universities or take up posts better than those of the parents, their approval of what were 'new schools' becomes clearer. Teachers in these schools feel under a challenge, feel the urge to produce good results to justify their existence, but not so strongly as in 1959. Hardly less important is the cumulative effect of school generations sad to leave comprehensive schools where pupils have enjoyed school life, including the variety of out-of-class activities, so generously provided where the staff is enthusiastic. In other ways, too, the quality of the teachers impresses parents.

These representative comments from our members speak for themselves:

(*a*) 'At the beginning, many parents were disappointed and suspicious. Recently they have become more enthusiastic. The reputation of the school has grown and misgivings are decreasing.'

(*b*) 'Where other members of the family have previously attended this school, the majority of parents then prefer to send younger talented children here.'

(*c*) 'The feeling against us is definitely lessening as time goes on.'

The change in attitude is most noticeable in areas where parents could send their children to a grammar school under the same authority or to a grammar school under a neighbouring authority.

(*d*) 'We have many able children who *chose* to come to us.'

(*e*) 'As the school's reputation grows, the number of parents putting the school as their first choice increases.'

(*f*) 'As sixth-form courses have been built up, so the percentage of parents sending children outside the area has dropped.'

(*g*) 'We have been considerably oversubscribed. This may be because our academic record is good.'

(*h*) 'Objections in early years are changing to overt applications for places.'

(*i*) 'Far more parents want their children to come to us than we can accommodate.'

(*j*) 'Nearly 100 parents from outside our zone opted for this school rather than for a selective school.'

When one counts how many schools are too young to provide such evidence, how many schools are in areas where there can be no parents' choice, this assembly of comments commands respect.

WHO GAINS? WHO LOSES?

Neutral observers, in the sense of those who are not pre-judiced either for or against the comprehensive system, have watched closely the progress made by the most able child-ren. Objectors have stated that the comprehensive school cannot do as well for the really able children as does a grammar school. Teachers in comprehensive schools are rather sensitive about this charge, which they think unjust. Only in a minority of schools is there an express stream, a 'fast form', with a four-year course to 'O' Levels. At our conferences, several teachers objected to this in principle and claimed that the high-flier is better for being in a five-year course. One correspondent told us that very good pupils were 'slowed down for a time in their school subjects'. He continued: 'When the really able pupil gets into the sixth form he gets all the individual attention—and some-times more—that he would in a grammar school.' 'Parents' choice' means different things in different parts of England. In one part the Local Authority may have both grammar schools and comprehensive schools, in another part, where all the former grammar schools have become comprehensive schools, a parent may ask if his child can be admitted to a grammar school under a neighbouring Authority. The attitude of a parent to comprehensive schools may depend on whether there are left any grammar schools to which his son can go. Grammar schools which receive such children may have a higher than normal number of 'high-fliers' and comprehensive schools a correspondingly lower than normal number. From many schools we hear that the best group of those taking 'O' Levels is not quite as good as the best group of a grammar school which does not lose very able children to neighbouring schools.

The danger in making a comparison between comprehensive school and grammar school is in the assumption that there is uniformity in both. One member, now in a young and vigorous comprehensive school, writes:

The comprehensive school is unlikely to better the standards of the better grammar schools but is likely to equal them, and gives a far wider general education to its pupils than the average grammar school can hope to.

A warning against considering separately the pupils who would have been selected by 11-plus machinery is contained in another review:

It is the considered opinion of the former grammar school masters here that the comprehensive system does not lead to 'dilution' as is so frequently claimed. Children of high ability, we are sure, would not suffer by the ending of segregation, but a great many other children would benefit by their presence within the school.

Our members are agreed that the boys who were not 'high-fliers' and who would otherwise be in the C form or D form of a grammar school do better in a comprehensive school. A representative opinion is:

The lower range of selective intake are the main beneficiaries of the greater range of subjects available from the fourth year onwards and of the possibility of combining G.C.E. and C.S.E. subjects. There is a greater tendency for teachers to adjust techniques and academic objectives more closely to the abilities of the pupils than in grammar schools.

All our members think that a comprehensive school can do more for the more able secondary modern boy who would not have been selected by 11-plus machinery.

For the boy of the A form in a modern school, the comprehensive school is God's gift. He has the full opportunity of a G.C.E. course to 'A' Level should he show any talent for it. If not, he has a wider range of 'O' Levels and C.S.E.

One school reports that one-third of its sixth-form pupils are from modern school material and that one-third of the university places were won by such pupils. Another school sending ten to universities in one year noted that only one of the ten had 'passed his 11-plus'.

Many members comment on their experience of the way the organization of a comprehensive school brings forward and brings on those pupils who might not have been considered as suitable for 'O' Levels. The experience of one school, we are told, 'seems to confirm that there are a large number of pupils within our schools capable of higher standards than is generally realized'. This is supplemented by opinion from another school:

As far as the average and less able children are concerned, they receive a much fairer crack of any whips that are going and can take advantage of all the non-academic courses open to them, which might not be the case in a modern school.
The children are far better catered for both academically and pastorally in the comprehensive than in the modern school and, I suspect, considerably better than in very many grammar schools.

TEACHING IN COMPREHENSIVE SCHOOLS

Very few people are aware of the extent of the growth of comprehensive schools. *Statistics of Education* gave the figures for January 1964 as 231 such schools in England and Wales, with 123 schools having fewer than 1,000 pupils. For January 1965 the figure was 262. The total number of pupils in comprehensive schools in January 1964 was 199,000. According to official sources, the position in January 1965 was as follows:

Type of secondary school	Pupils		Teachers*
	Number	Percentage of total	
Modern	1,555,132	55·2	78,576
Grammar	718,705	25·5	41,862
Technical	84,587	3·0	4,965
Bilateral and multilateral	66,166	2·3	3,549
Comprehensive	*239,619*	*8·5*	*13,364*
All secondary schools	2,819,054	100·0	150,678

* Including part-time teachers at their full-time equivalent value.

The Minister of State for Education was reported in *The Guardian* of 1 October 1966 as saying,

Provisional figures for January, 1966 showed that there were now 342 maintained comprehensive schools. This meant that about one in 10 of the secondary school population—or well over a quarter of a million—were already in comprehensives.

3

VARIETY AND SIZE

If the British Broadcasting Corporation wishes to put into a television programme something about comprehensive schools, it seems that it is always one of the show-piece palaces in or around London that provides the pictures. It is partly because of the writers, amateur and professional, in daily papers that the public attaches the description 'vast' or 'monstrous' to comprehensive schools. When it is decided to create a comprehensive school, too many adults assume immediately that the local Council will imitate the very largest member of a family of comprehensive schools. It is a growing family, with the eldest still under 21 and the youngest still to celebrate its first birthday. If 'average' figures are of interest, one would have to say that the average age was about seven. The English schools from which our members have supplied us with information show this pattern of establishment

Year	1952	1953	1954	1955	1956	1957	1958
Number	3	1	4	5	7	1	4
Year	1959	1960	1961	1962	1963	1964	1965
Number	5	5	4	3	4	7	1

In the summer of 1965, fifteen of these schools were not able to enter in the C.S.E. or in the G.C.E. any candidates who had come up from the first year in a comprehensive school. Not until 1968 will these fifteen young schools present any 'A' Level candidates who have come right through a comprehensive school.

The London area is the place for single-sex comprehensive schools and there you could reasonably talk in terms of a typical boys' school or a typical girls' school. Outside Greater London there are very few boys' schools that are comprehensive schools. Every Jack must have his Jill: there can be a single-sex comprehensive school only in an area with sufficient children to separate into boys' school and girls' school, without causing many children to travel some distance. In theory, the Head of a mixed comprehensive school can be a woman: it happens so in practice, too.

The following section (on buildings) will show how unjustified is the association of brand new buildings with new comprehensive schools. It should be stated over and over again that to give a new school the good start it deserves it needs buildings planned for its needs. When a Local Authority is impatient to have comprehensive schools it can do little but survey the buildings that already exist and conscript them. Quite a few comprehensive schools began life only because of something like a take-over bid. New, unoccupied buildings, planned for the needs of a different kind of school were taken over, filled with children and the name 'comprehensive' applied. One comprehensive school took over buildings designed for three separate secondary modern schools; several schools took over buildings designed for bilateral schools. Change of power can bring change of plans; buildings on one site can in the course of a few years be used for different schools. One school that was a secondary modern school in 1961 became comprehensive in 1963, with a spell as a bilateral school in between.

The origins of a comprehensive school affect the composition of the teaching staff. A completely new school, one not taking over buildings of any other form of secondary education, not absorbing pupils and staff from any other kind of secondary school, can build its own traditions without having to modify or abandon competing traditions.

About one in six comprehensive schools have absorbed or added to an existing grammar school. But of the small (under 650) schools in rural areas, four out of five have a grammar school nucleus. Such schools continue the tradition of working for external examinations, the tradition of service in out-of-class activities, the attitude (in scholars as well as in staff) that sixth-form education is the normal completion of school life. Beneficial traditions came, too, with the staff of central schools, intermediate schools, selective high schools, technical schools. (At least one example of each kind has been absorbed.) Schools which absorb teaching staff must be ready to see a bigger than usual rate of leaving as those who do not like the change find other schools to work in. In the completely new school, where all the teaching staff are chosen from applicants who choose to teach in a comprehensive school, there is a higher degree of devotion to the comprehensive principle.

Parents contribute to the variety within the present range of comprehensive schools. Parents in one area may have much greater opportunity to send their boys and girls to another school, a direct grant school or a boarding school or a neighbouring grammar school which will accept the children. There is no means by which planners and administrators can ensure that a comprehensive school can get its fair share of the ablest children. In the school which has a larger than usual proportion of pupils leaving as soon as the law allows, what is reflected is not the quality of the teaching staff but the outlook and attitude of parents.

If comprehensive schools are to be divided into categories, sense and justice require some classification like this:

(A) Boys' schools in London. Size: considerable variety. Most have now nine years' experience behind them. Neighbourhood schools, strongly affected by the homes in their area. Modern school in origin, as a rule.

(B) Coventry schools, single-sex and mixed. Size: above average. Well established with ten years' experience. Planned with insight and foresight, provided with excellent buildings designed to give expression to aims. Here the House system is strongly established; a pupil lives and learns in his own House. Range in size: 1,300–1,900 pupils.

(C) New area schools, sections of large cities or overspill towns where the schools were planned along with the houses, so that there are purpose-built schools without taking over existing buildings and incorporating staffs of schools absorbed. Very much neighbourhood schools for new housing estates. Size: usually 10-form entry or 12-form entry.

(D) In larger cities, schools which represent the early stages of plans to provide comprehensive education throughout (Leeds, Sheffield, Bradford, Bristol, for example). Whether existing buildings are used or specially-planned buildings erected seems to depend on the urgency attached to making a start. The catchment area may be arbitrary rather than natural, subject to change as the total plan proceeds. Whatever attempts are made to make the pupils of one school comparable in range and class with those of others in the same city, sooner rather than later the schools of one city become graded in public esteem. Most of these schools are in the growing stage. Some have pupils who came in as all-age schools were closed, pupils who did not begin in the first year. Some are affected by a Local Authority's change of attitude to direct grant schools. The range in size is from 8-form entry to 12-form entry.

(E) Area-unit schools (not serving a section of a city), e.g. a school that takes in all the children of one unit of local government such as an Urban District Council. Usually the nucleus is an existing grammar school, a small one unlikely to expand because it serves a static, scattered population in a rural area with the small town as its natural centre. Often Authorities do their best to provide good buildings but sometimes there is no alternative to accepting existing buildings as a base. The more rural the catchment area is, the less likelihood of the school exceeding 700 pupils. Making a comprehensive unit may have been the best way of dealing with small secondary modern schools which were inadequate. The combined effect of the grammar school nucleus and the lack of variety in opportunities for employment seem to lead to a higher percentage of pupils completing the fifth year.

Reflection on the major types and the variety within each type leads one to think that no two comprehensive schools are exactly alike.

SIZE

Before one passes judgement on the size of a school, one should find out if it is fully grown or not. In Table 1 we attempt to show the range in size of English comprehensive schools, excluding those which began after 1962.

Another way of looking at numbers is in terms of intake (Table 2). In this way, figures from growing schools can be included. It must not be assumed that a multiple of 11 means an 11-form entry; provision for remedial classes, small ones, will often turn a school with 11 × 30 into a 12-form entry school.

The ratio of total numbers to first-year intake numbers depends on the area (kind of home, chances of employment), the prestige of the school, and the policy of the school. It is assumed that instances where that figure was for special reasons higher than the average for that school will balance instances where special reasons accounted for a lower than average intake in 1965. Table 3 is limited to 36 English schools, excluding those whose certain further growth

TABLE 1. *Numbers in English Schools, September 1965*

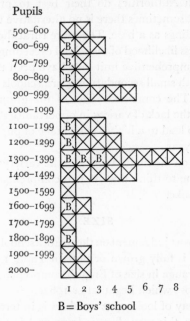

B = Boys' school

makes the total numbers for September 1965 unrepresentative. (One school, with an intake of 84 and a total of 600, appears to have a ratio of 7; it has been excluded because it used to have an intake of 110.)

OPTIMUM SIZE

Some theorists feel on very safe ground when they give figures for the optimum size of a comprehensive school. Though they seem to know all the reasons for their figures, they lack one essential—experience of working in a comprehensive school. The individual teacher may accept that there is an optimum size to the school in which he works

TABLE 2. *Intake for first year in multiples of 30.*
English schools, September 1965

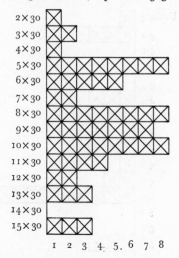

but he will be ready to point out that this is not necessarily the optimum size of any other school. It is difficult to find a man with sufficient experience in two schools of different sizes to be able to make comparisons. In a sample vote at one conference of sixteen members six thought that an 8-form entry was the best, eight preferred a 10-form entry, two wanted something larger. For a full appreciation of these choices one would need to know the size of the school in which each voter worked. If you ask a number of people, 'Which would you prefer—to be slimmer or to be fatter?', the answers mean little without photographs (or figures) of the people providing the answers.

Some teachers have been in schools which have undergone a change in size. A member in a very large school tells us, 'If the school were the size intended (12-form

TABLE 3. *September 1965 total as multiple of September 1965 intake*

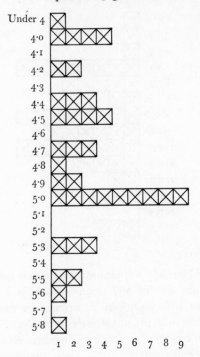

entry) life would be easier for the staff but I think it might now feel a bit empty if we lost a large number of children'. A man who had experienced a drop of 300 in numbers preferred the smaller number because of reduction of pressure. Clearly, once a school is built, the optimum number for that school is almost settled by the accommodation and facilities. It seems that buildings well designed for pupils' movements can take without discomfort a larger number of pupils than buildings intended for the same number but which have poor design or uneconomical layout.

Many schools complain of insufficient accommodation but very few of too many children. 'A fully effective comprehensive school could be operated with far fewer pupils—certainly 1,400, perhaps, 1,000' is an unusual comment from a school of 2,000. One school of 1,250 considers itself 'quite big enough'; another of 1,325 is considered 'about right' but it is admitted that it 'could be rather smaller'.

Members who have found that a comprehensive school can work well with 1,250 pupils or more are often ready to accept more. 'If we had to make a choice, we should prefer the school to be slightly larger rather than slightly smaller' comes from a school of 1,450. Another comment is: 'The school functions fairly smoothly at its present size (1,150) and would probably function no less smoothly if it were larger.' A member in a school of 950 thinks it 'not as big as it ought to be'.

Very often what affects the judgement of our members on optimum size is the provision of courses. 'In order to provide a diversification of education, for example in the technical groups, sufficient numbers must exist to make that variety justifiable both in terms of outlay and of employment of specialist staff' is behind a wish for 1,500 in a school, where the present numbers are short of 1,200. Experience of 800 leads to the view '1,000–1,100 would allow more flexibility in terms of options and choice of courses'. Six hundred is 'too small to allow full development' because the staff 'realize the possibilities that would come with another 400 pupils (and 20 staff)'. In a school just short of a thousand, the staff are 'emphatic that this is too small'. An 8-form entry, it is thought, is needed 'to arrange a better choice of subjects in the fourth year and the fifth year'.

From the beginning, decisions at the planning stage about the number of pupils for a comprehensive school have been based on theories about the size of the sixth form. The oldest comprehensive schools in cities were planned on the assumption that a school of 1,800–2,000 was necessary to make sure of a good sixth form. Now there are much smaller schools which have a bigger sixth form than the first large schools produced. The increase in the number of pupils voluntarily staying on after reaching school-leaving age, together with the wider conception of sixth-form work, has made it necessary to revise the original maximum figure of theorists. There is no general agreement about how large a sixth form should be. Outsiders may talk in terms of 100 but teachers see the advantages of 200 in the two (or three) years of the sixth to provide for all reasonable demands. Desire for a strong sixth form weighs strongly in small schools (600–750). One school (of 950) feels the need of 1,250 'to provide adequate sixth-form numbers in order to be economic, especially in Arts subjects'.

The possibility of treating the sixth form as a separate unit may make it necessary to reconsider the older ideas about optimum size. Where the House system is the basis of organization and where buildings are House-planned, it is thought that 200–230 is the optimum size of a House. If the House is of the right size, the size of the school is less important. With four Houses, the school could be 800–920; with six Houses, 1,200–1,320. With a separate sixth form— taken out of the House system or made a separate House, then the optimum size could be larger still.

Another approach is suggested by schools where the year system or the horizontal division is preferred to the House system. If you divide the school into Lower School, Middle School, Upper School, with a Head for each division, then it is possible for one man to know as individuals up to 600 pupils in one of the three sections. If that is so, it seems that a school could go up to 1,800 in size.

24

One mistake too often made is to work out figures for the ideal size of a comprehensive school as if the quality of the intake were constant. A member in London thinks his school can be 'only quasi-comprehensive; 18% go to grammar schools'. Other members claim that the intake must be bigger because of this loss. This is a problem in the Home Counties, too, where parents can opt out of the comprehensive system. There can be dissatisfaction with a total of 1,350 pupils because with the loss of really able pupils the sixth is limited to 70. Another member makes his figure of 1,400 as the optimum size conditional—'if fully comprehensive'.

In some ways the small rural school has its own optimum size. Its sixth form, still small, may be bigger in proportion, yet with small numbers it is harder to do justice to pupils needing unusual combinations of subjects. Teachers who grow up in a small school can think of 600 or 700 as 'just right' but teachers who move from a larger school wish for larger numbers. To increase the number of pupils in rural areas would be to increase the problems of travel. It is harder to provide specially trained staff for remedial classes for the backward. Small numbers result in one man having double responsibility—Head of Department and Head of House—and in fewer teachers having opportunity to take responsibility. Yet there may be no workable alternative to operating small schools. If the aim is to bring together in one school all the children of one local government area, all the children of one housing estate, all the children of one valley, all the children in an area formed by river boundaries ('where the catchment area is clearly defined geographically') the starting point and the limiting factor is the number of children, not theories about size. Where you accept the natural area, you accept a certain size.

4

BUILDINGS

The man in the street seems to think that a comprehensive school consists of imposing brand-new buildings, incorporating the latest designs and devised by forward-looking architects. In fact, about half of the comprehensive schools have new, specially-designed buildings. Yet every comprehensive school deserves a fair and proper start, toward which made-to-measure buildings can contribute so much. We need no public opinion poll experts to tell us that the reputation of a school depends a good deal on the buildings provided. Parents naturally take pride in a new school which they have seen grow out of the ground. When administrators decide to take existing buildings and alter a little here and there or add a section or a few rooms, the public understandably is slow to accept it as a 'new school'. Children, too, are more likely to take pride in a full set of new buildings 'designed and built for us'. Teachers who come for interview are both human and sensible if they allow the appearance and design of the buildings to influence their decision whether or not to accept a post there. For the man at any rate, a school is his workplace and that for more years than those spent in the school by a pupil. For those parents who have an alternative school to which to send their children—a grammar school a little further away or a modern school nearer home—the quality of the buildings provided is an important factor.

Of 50 comprehensive schools in England of which we have full details, 24 began with new, specially-designed buildings. These include the schools of Coventry that opened in 1954 and 1955, schools in which the teachers seem to be particularly happy. London provided several examples of completely new buildings between 1953 and 1960. New estates and overspill towns often incorporate new schools in the overall building plan. Industrial cities—Bristol, Bradford, Hull, Leeds, Liverpool, Sheffield—have between 1957 and 1964 included new-building schools in their programme of re-organizing education on a comprehensive basis. Schools with purpose-built buildings in rural areas are the exception rather than the rule but between 1962 and 1964 such schools were provided in Cumberland, Devon and Nottinghamshire.

Of the 50 comprehensive schools, seven have been based on an existing grammar school. There is no standard pattern, as these summaries will show:

(1) Grammar school from 1907, extended in 1956 to make it comprehensive.

(2) Grammar school of 200 pupils, extended in 1962 to bring in children from 10 all-age schools.

(3) Grammar school with 2-form entry from 1960 until 1963, when it became a 5-form entry comprehensive school.

(4) Grammar school from 1950, adding later all the children of secondary school age from many all-age schools.

(5) Grammar school of old foundation but with only 170 pupils. It became comprehensive in 1963, taking children of secondary-school age from all-age schools. Old buildings of the grammar school, half a mile away from the present comprehensive school buildings, were used until 1965, when the second phase of building was completed.

(6) Grammar school opened in 1928, amalgamated in 1957 with secondary modern school. Buildings of the grammar school extended in stages.

(7) Grammar school with a technical adjunct became in 1956 the base of a comprehensive school, with new sections added.

We realize that adding sections or blocks to an existing grammar school has been common during the last 15 years and that the growth in numbers of those educated in grammar schools has meant that it is now exceptional for a grammar school more than 30 years old to contain all its pupils in a single building.

A grammar school, even a small one, is likely to make a better base for a comprehensive school, formed by amalgamation, than is a modern school. A grammar school so absorbed carries with it its physical provision for a sixth form—a library, laboratories for advanced work in Chemistry, Physics, Biology. A modern school may not have one room devoted entirely to holding and displaying works of reference and providing accommodation for students. A modern school will have some provision for science, but no separate laboratories for the separate sciences. It is unlikely to have preparation rooms for work in laboratories. Until these deficiencies are made good, a comprehensive school based on a modern school is under a handicap.

Of the 50 comprehensive schools in England that we are here considering, 15 began with buildings designed for other forms of secondary education. Variety is to be expected.

(8) Opened new in 1958 but in buildings initially intended for three secondary schools on the same campus. 'The sprawling nature of the campus', it is felt, operates against out-of-class activities as well as causing excessive movement of staff or pupils.

(9) Opened in 1956 in buildings intended for a bilateral school. 'No proper acid store.' Cupboard and storage space inadequate. Still no prefects' room or study room. No music rooms. No Advanced Science laboratories.

(10) Opened in 1960 by amalgamating a central school (1875 buildings) and a senior modern school (1905 buildings) half a mile apart. Lower school in one of the two sets of buildings. Specialist rooms inadequate. P.T. and library facilities most inadequate.

(11) Opened in 1948 in a 1938 building for two single-sex schools. 'Only during the past year, 16 years after the school became comprehensive, has the new extension of library, science labs, technical drawing rooms, geography room, been opened and a second gymnasium built.'

(12) Opened in 1964. Amalgamation of two single-sex modern schools. 'Lab space is stretched to the limit, only four labs', where eleven would be reasonable.

(13) Opened 1954 in buildings designed for a modern school. 'The science labs in particular are inadequately equipped with ancillary rooms. Chemistry, Biology, and Physics labs have only one preparation room (shared) and small storage cupboards. Now plans are afoot for provision of new Advanced laboratories by conversion of classrooms.'

It seems to take years to correct initial defects and deficiencies. Some schools are still in the midst of alterations and extensions. This is understandable with grammar schools dealing with a recent increase in numbers. One would expect a comprehensive school opening in 1963 to start with sufficient rooms, yet one reports: 'Existing buildings supplemented by new block opened September 1965 and 14 temporary huts. Eventual plans for another building.' An even younger school has its difficulties. 'New school opened in 1965 in premises of new modern school. In January 1966 work will start on building the rest of the school and on alterations to existing premises.'

Other schools are awaiting the final stage of building. One school opened in April 1956 and occupied the Stage 1 buildings in autumn 1956. Stage 2 was added and occupied in autumn 1960. Stage 3 will be built in 1966 and occupied in 1967. Possibly the teachers with a firm date for the final stage are happier than those in the school (opened in 1963) in buildings 'to develop on a 3-phase plan, phases 1 and 2 being complete. There is no date for phase 3.'

Waiting for improvements has its compensations. The delay may provide an opportunity to incorporate features that would not have been considered earlier. One London school expects, at Easter 1966, gymnasium and games hall, 'believed to be the first of its kind in London. There will also be a specially-designed climbing wall for instructing in the basic principles of rock-climbing.' In addition there are to be four marking-rooms, belated recognition of the teachers' need for somewhere to work as well as for somewhere to chat. Nor are the planners afraid to go higher: the new teaching block will have seven floors and will be equipped with lifts, each to carry 30. It will be interesting to watch this venture that avoids long stretches of corridor and countless stairs and that admits that boys are capable of operating lifts on weekdays, not on Saturdays only.

One big advantage of having additions and extensions after the opening of a school is that there is much more opportunity for practising teachers to make constructive criticism as well as to point out defects. A set of school buildings can bring a legacy of irritation and frustration. The architects enjoy trying out their schemes; the teachers have to live with the experiments and work in them. Too often a new comprehensive school has been erected before a Headmaster is appointed. The plans have been drawn up by theorists and approved by administrators, possibly not seen at all by anyone who has worked with the children of these ages and in these numbers. If a Headmaster has been appointed before the plans reach their final form, he can watch the interests of pupils, of teachers in general, and of specialists. It would seem that there is an opening for an Advisory Committee or Panel of teachers experienced in comprehensive schools who could 'vet' the plans for a new school which has no Headmaster to scrutinize and suggest. (One city set up such an Advisory Panel in 1966.)

A decline in numbers is rare and happens because the local authority has re-deployed the children of school age in its area. There is no question that the general movement of numbers of pupils is upward, almost entirely because of the increase in numbers of children voluntarily staying on after they have reached the age when they may leave. The C.S.E. will make fifth forms bigger. As more of the pupils are brought into the G.C.E. sphere, more stay on, to make sixth forms still bigger. There is more and more need for rooms larger than normal classrooms. If the lower sixth numbers 50, it is only on the half-a-loaf principle that you can offer them a classroom for their common room. Thus, plans which are on the drawing-board now but which are not forward-looking in this respect will certainly prove inadequate before the buildings themselves have been occupied for a year.

Those who study and approve plans for new schools and for extensions to existing schools may feel quite pleased to see rooms earmarked for storage space. They may have little idea how much has to be stored. Only an experienced Head of Science Department will know what apparatus, what materials have to be accessible on shelves or kept out of reach. A large department such as Mathematics, English, Modern Languages, needs a room with generous shelving to hold all the text-books at the times for collecting and issuing. The syllabuses drawn up for C.S.E. often call for more books as well as different books and so the need grows for more storage space. G.C.E. set books, after spending one or two years in circulation, need a place in which to hibernate until they become wanted again. Some schools prefer to have the teaching aids providing pictures or sound in a central store-place from which they can be collected and to which they can be returned. The more active a school is in drama and music, the more material there is. A busy school has masses of material which needs to be kept where staff can have access to it and where it is safely away from children. Particularly do they need to be protected against that arch-enemy of schools—dust.

Lack of imagination or zeal for economy leaves the teachers of some schools chafing several times each school day— when pupils are on the move. One school complains of a path three feet wide over which 500 children have to pass in five minutes. What chance is there for grass to survive at the side of the 36 inches? A campus site may have attractive buildings spaciously separated with up to 200 yards between blocks. This looks delightful on special occasions or if an aerial photograph is wanted, but on ordinary days 'when it rains we all get wet changing classrooms'. Even blocks not so far apart call for covered ways as a defence against rain and snow. In other schools teaching time is lost and noise is increased because passages, corridors, staircases are too narrow. Movement areas should be planned with maximum demands in view.

In many schools, wet intervals, wet dinner-times are trying periods. Where are children to go? If there were covered areas, some children could be outside and they could have exercise instead of sitting inside. For Physical Education such covered ways would be very useful, almost an extension of a gymnasium, so that a class could be split into different groups for different forms of activity.

In many schools where the teachers wish to make fuller use of teaching aids the enthusiasts are handicapped by inadequate accommodation. A school with fewer than a thousand pupils deserves to have a room equipped and designed for television. One school goes further than making use of televised programmes for schools; it has its own closed-circuit system. For this, a number of rooms with receiving apparatus must be available. It is not good enough to have a classroom designated as Projector Room or TV Room if a class has to be turned out before it can be used. It is sensible and economical to regard such rooms as special-purpose rooms to which pupils go, just as boys go to a Wood-work Room or girls to a Domestic Science Room. To serve the need fully, there should be several specially-equipped rooms, one of which could be a Visual Aid Theatre, with seats banked so that the back row is as good a place as any-where else. Planners should be anticipating colour-television, the day when Schools Programmes in colour increase the need for receiving rooms.

Striking changes in teaching methods are creating other needs. Possibly at first the official view will be that a langu-age laboratory is justified only in a very large school. Such an attitude should be opposed, strongly and constantly. If learning to speak modern languages is worth doing and if a language laboratory is as good a means to this end as has yet been devised, then the size of the school should not be used as evidence against the school's demand. Would it not be just as reasonable to claim that for a very large school one language laboratory is not enough? Our teachers in comprehensive schools include many forward-looking en-thusiasts and experimenters, far from being conservative and satisfied with talk-and-chalk teaching. Planners must be progressive to cater for progressive teachers. One school has a Mathematics laboratory on the way; other pioneers in the new approach to Mathematics will be less able to do their work thoroughly until proper provision is made.

The Music teaching in schools today is often quite

different from what it was when the earliest comprehensive schools were opened. With the increased interest in instruments of all kinds, accompanied by an increasing use of peripatetic teachers in school-time, it is not now sufficient to plan and equip for Music in terms of single or double classes engaged in singing. Rooms for small groups—'division rooms' in the language of architects—are needed in large numbers in up-to-date schools and possibly no subject has a stronger claim for them than Music. Music on the time-table may involve 'cellists in one room, players of wood-wind instruments in a second room, and those learning to play brass instruments in a third. The need for care in siting such rooms should be obvious.

Recognition is needed for adventurous teaching in Domestic Science and in Commercial Studies. The worth of a Domestic Science House (or Flat) has been proved in several schools. Such provision deserves to be a standard feature of plans for mixed schools. Two schools have succeeded in getting a Model Office provided so that the Commercial Course leads to doing work in the like-real-life office just as other subjects lead to practical work in a specially-equipped workshop. Another activity for which demand is increasing is Technical Drawing (Geometrical and Engineering Drawing). With so much work to be done on half-imperial sheets, there is justification for a room devoted to this subject, with special desks for the large drawing boards, too large for ordinary desks.

A school which believes in making parents welcome deserves an attractive room in which to receive them. However well a Medical Room is equipped, that is insufficient if there is no comfortable place—something much better than a chair in a corridor—in which parents can wait and feel at ease. There is much more frequent need for an Interview Room in which parents can talk things over with the Housemaster or the Head of Middle School or any others whose concern is pastoral care.

Teachers are aware of the growing need for the right conditions for study and leisure of the sixth form. If these pupils are to be treated as responsible young men and women, the building programme must include a Common Room as a social need. One school has the right idea in asking for a sixth-form House which includes a room in which the students can do light cooking. It is easier to train students in habits of work if the buildings provide good facilities for private study. An ordinary classroom with ordinary desks is not good enough. One feature of some of the new sixth-form units is worth notice—a room partitioned into study-units, each providing shelves for books and bags, ample writing surface, the partition preventing one from being distracted by one's neighbour. For the Staff as well as for the students there is a need for rooms in which a sixth-form tutor-group can meet.

To this list of items which our members consider would improve teaching and learning in a comprehensive school—almost all of which are already permanent features of one or other of the best equipped schools of 1966—we should add one which is likely to be more expensive. Some comprehensive schools already have their own swimming pool or swimming bath; many others place this high in their list of requests. For this must be of benefit to pupils of all ages. In some areas the money-saving advice of 'Use the public baths' is laughable. Even if a special coach was used to take a class to the baths and bring it back, it would often be impossible to provide a visit to the baths without taking more than one lesson. It is hard to calculate the double value to a school of having its own baths—value in teaching time and value when lessons are over.

35

5

CARE FOR THE INDIVIDUAL

'A new pupil must feel lost when he joins a comprehensive school' is what we have all heard and often. The same generalization could be made of a new pupil joining a large grammar school. In both instances, the word 'must' should be challenged. Our members in comprehensive schools would change 'must' to 'need not'. This section brings together examples of different approaches and different methods made and used in schools of 1965.

The 11-year-old will come from a primary school where Jack Robinson is a person of some importance in a little world. If he moves to another kind of school—even a modern school—the new school will be bigger, with unfamiliar buildings, with every adult a stranger. Instead of being one of the oldest in the school he is now one of the youngest. The oldest ones in this new school are much bigger, much older. In a short time, if he is a normal boy, he makes friends with one or two boys of the same age in the same position of 'new boy'. He gets over the initial nervousness about the unknown and begins to feel at home. Whereas this period of growing acceptance is left to natural processes in some schools, in comprehensive schools very much indeed is done to help Jack Robinson to feel he belongs.

The ideal is that from the start he is known and cared for as an individual. 'The problems of size are organizational; they can all be overcome by a well-organized system' is one member's summary. The school can set up a framework within which friendly adults readily play their part in making Jack Robinson know that he matters to someone.

36

This personal care is to operate for more than a few weeks; it must be continuous and ever available. When that is done, the big school is not unwieldy or oppressive, the individual is neither neglected nor submerged. The large numbers allow unorganized natural processes to operate, giving the individual boy a greater chance of finding a congenial companion either of very similar nature and tastes or of strikingly different ways and temperament. 'The lonely "misfit" that one frequently sees at the small grammar school and the extremely late developer at the modern school who is isolated because of his limitations are rarely found at the comprehensive school. There is little chance of being the odd man out.'

Men on the scene of action are sure that adults worry more about size than children do. 'Pupils do not appear to be conscious of the size of the school, except for the first year during their first week or two.' From a school of just over 900: 'In eight years I can recall only one pupil who had obviously not been able to adjust to the size.' From a school twice as big:

The size of the school has the initial effect of causing children to feel rather bewildered and leads to some of them, especially the less intelligent, getting lost themselves and losing their belongings. It takes rather longer to overcome this problem than it would in a smaller school. Within a week or two, most children grow quite accustomed to the size of the school and the large numbers in the building and settle down very well.

How essential is it to feeling at home in a community that the whole community should be able to assemble in one place, to be seen as a whole and to be felt as a whole? One does not expect a university to have a hall which will hold everyone. If one studies figures for the size of grammar schools as found in the 'Situations Vacant' pages of educational journals it is clear that some schools survive without a 'full assembly'. Teachers in comprehensive schools are surprised that others ask if the whole school can meet in an assembly hall. If there were a hall big enough (and it would be unreasonable to ask for one big enough) it would not be wise to attempt regularly to assemble 1,300 pupils. Some grammar schools know the benefits of having a separate Junior Assembly; very many grammar schools hold successful House Assemblies once a week. So in comprehensive schools, daily assemblies are held according to the halls available; sometimes the House is the community, sometimes the intake of two years (Junior School) or of three or more years.

'The Headmaster cannot possibly know every pupil in his school' is another trite criticism. This *is* a problem of size and does not depend on the type-label of the school. There is an upper limit to the number of pupils one man can know even superficially. Above that limit, in modern school or grammar school, the work of getting to know pupils as individuals has to be shared. There are grammar schools which have a Head of the Junior School, some with a Head of the Middle School in addition. Such delegation, possibly an occasional feature in grammar schools in 1965, is the normal way of school life in comprehensive schools of over 500. The delegation is built into the organization; the administration and the provision of pastoral care almost merge.

English comprehensive schools of 1965 show much more widespread, more thorough, efforts to break down the big organization into small units than was to be seen in 1959. Here we have an example of the persistence of enlightened Headmasters in trying and trying again until they find a method right for the groups of children and groups of buildings which make up a comprehensive school. As new schools are opened, senior administrators are appointed who often start with the form of organization to which they are accustomed and of which they have experience. They are ready to change it if it does not work. We now have two major patterns (with scope for local variety), both based on the principle of double division. First, the boy comes under the control of the Headmaster's delegate—the Head of Junior School, the Head of Middle School, if the division of the school is by years, or the Housemaster, if the House organization is used. Because the horizontal group of Junior or Middle School can be so large, itself the size of some grammar schools, many schools prefer vertical division where the number of Houses can be in ratio to the total numbers to provide a House group of about 200. The second part of the double division breaks up the larger group into a number of small groups of about 30, a tutor-group under either House system or Year system. So first the individual boy is one of 30 carefully 'fathered' by one man. Secondly, he is made aware that his progress and welfare are the concern of a much more important man, the Head of School or the Head of House, the man who writes to his parents and answers their letters and interviews them.

This means that the Headmaster is behind his delegates and, seemingly, further remote from the pupil. The Head-master's influence is still felt; after all, the detailed organiza-tion of delegation which provides the pastoral care is the policy of the Headmaster. The ideals and principles of the Headmaster inspire and guide the men to whom he dele-gates authority. The organization can be compared with that of a big firm—Chairman, the Headmaster; Deputy Chairman, the Deputy Headmaster; Managing Directors, the Heads of School or Heads of House. In fact there have been created a number of Limited Area Headmasters under a Major Headmaster. Each of the former may have as much responsibility and as many duties as a Headmaster of a small school. The success and progress of the individual comprehensive school—indeed of the comprehensive concept —depends much more than is realized on the calibre of these chosen men. A Local Authority which believes in justice will see that their worth is recognized in the distribution of special allowances. It will give them proper conditions so that they can do their valuable work, by adapting the staffing ratio so that these key-men can be relieved more from teaching duties. Each Minor Headmaster needs a room of his own, where he can meet parents and see pupils.

The current of organization to provide thorough and continuous pastoral care is carrying away the old idea of the Form, that seemingly permanent feature of the streamed but unsetted grammar school. The practice of one group of pupils, one unit for registration, going together from lesson to lesson from Monday to Friday, is being abandoned. The group of 30 or so of which the new boy becomes a part is a *social* unit rather than a unit based on grades of intelligence and set programme of lessons. The man immediately in charge of the small group does much more for and means much more to the individual boy in a comprehensive school than does a Form-master of a grammar school form. The pastoral-care unit is becoming distinct from the class-teaching unit. There need be no regrets for the change. The newer system brings more adults to accept responsibility for and to take deep interest in the welfare of pupils. This is good for the adult as well as for the child. What is good for both is good for the school.

Division by years

At least three schools are almost driven to divide by years; each has an auxiliary building a quarter of a mile or half a mile or a mile away from the main buildings. The distant building becomes the Junior or Lower School.

Undoubtedly there the man on the spot in charge of it all is the equivalent of a Headmaster. Apart from those we note these arrangements:

Head of Junior (or Lower) School without corresponding Heads for the rest	5 schools
Head of Junior (or Lower) School along with Head of Senior (Upper) School	5 schools
Heads of Junior and Middle Schools without Head of Upper School	5 schools
Three separate Heads: Junior, Middle, Senior	3 schools

Sometimes where there is no Head of Upper School the Deputy Headmaster is overseer of the senior division; sometimes the Senior Master looks to the senior boys and the Senior Mistress looks to the senior girls. In the very large schools a threefold division can have one of its schools over 600; at one conference we heard of a middle division that exceeded 1,000. In such cases, there can be a Year Master or an Assistant Head of Middle School to assist with pastoral care.

An example of these intermediaries is found in a school rising to 1,500 pupils. The Head of Lower School looks after the pupils in their first two years (just over 550), concentrating on the second year and having an assistant concentrating on the first year. The Head of Middle School is in charge of the third and fourth years. He has two assistants for third year and three for the fourth year. One assistant looks after the leavers; two assistants see to those on a two-year course. Pupils in their fifth and later years come under the Deputy Headmaster and the Senior Mistress. To help them there are two assistants for the fifth year, while the sixth form come directly under seven tutors.

In the above school there are six Houses, retained for usefulness in organizing contests in games and athletics but

not an important feature of the organization. Here the
post of Housemaster is neither onerous nor significant.
In more than one school the House system has disappeared.
The man who told us 'We have killed the House system
in this school' seemed glad to see it go. Another school
carefully replaces the House system with a Year system.

Over the whole year-group are a Year-master and a Year-mistress.
They are responsible for all aspects of discipline, the welfare and
social activities of all the pupils in the year. This entails the
Year-staff dealing with disciplinary matters referred to them by
individual teachers, close liaison with the School Welfare Officer,
and the interviewing of parents about pupils' careers or personal
problems.

In this school it is maintained that the Year system is
better than the House system 'in not splitting the school up
into several entities, thus fostering the idea of belonging to
a school rather than to a House'.

At least two other schools preserve the House division
but make much more use of the Year-master for tutorial
care of around 240 pupils. The Year-master works through
the tutors who have their groups of around 30. The work
of tutors will be more clearly seen in the section considering
the House system.

Division by houses

Buildings can encourage or prevent the efficient working
of the House system. At one extreme, 'The building is
impossible to split down into small friendly units'. At the
other extreme, the fortunate school designed and delivered
to give a real home to a real House. An example of how a
very large school can use the House system as a corrective
of size is seen in this account of a school with 2,000 pupils.

The school is divided into eight Houses which are located in four
House blocks, two Houses to a block. Each House has its own

assembly hall and staffroom. The two assembly halls can be joined to make one large hall for lunch, House-parties and dances. Approximately 250 children live in a House. They assemble as a unit in the House, and eat as a House. Milk is served in the House. They are looked after by some 12 or 13 staff and so in many ways they seem like a small school within a school. The middle-ability groups are taught in House sets, and as far as the time-table allows by staff from either their own or an adjacent House. The younger children have a strong sense of belonging to a House and enjoy keen competition for trophies.

Where pupils can 'live in a House,' where 'our House' can mean 'the building we and we alone use and enjoy', the House system can work wonders. Some accept that strong loyalty to a House can mean less loyalty to a school. Its advocates certainly sing its praises.

It places the individual. It gives him a meaningful existence. An Upper, Middle and Lower School does not do this. Housemaster and tutor try to remove the problems which hinder the full development of the child. It places Housemaster and tutor in a unique relationship with a boy. They come to know a boy as a boy—his home life, his interests, his weaknesses.

Another supporter writes:

It provides a social unit of reasonable size, within which individual pupils can find dignity, security and a sense of meaningful 'belonging' to a group having a corporate identity and demanding loyalty from its members.

School life can have many pleasures for the boy whose nature leads him to become 'House-minded'. The social atmosphere can be a source of delight. At break, the House building gives him somewhere to go while he enjoys what he has bought from the House tuck-shop. Where one kitchen serves two Houses, the food comes hot to the boy as he sits with his House-mates. After dinner the House-room or the House grounds give space where he can play without infringing any rules. After school, he can do homework in a room in the House, or attend a meeting of a House society, or play in a House team competing with some other House. Is it surprising to learn that in such schools 'the community spirit is terrific' or 'Boys are the friendliest to masters that I have ever met'? And as he grows older, the individual can do more to serve the House and become one of its leaders, as monitor, or prefect, or team-captain. If loyalty has value and brings benefits, House loyalty is good.

Though special buildings may be highly desirable, the House system can work without them, but there will be more artificiality about the House system. The adults will find themselves working harder to achieve the same results. House loyalty may be a slower growth where there is no physical House. Special buildings or ordinary buildings, the experience of our members is that a good House shows its power in better behaviour from happier boys who develop talent and personality in a rewarding way.

There is some truth in the saying that 'A House is as good as its Housemaster'. Just as he should be inspired, so he should be able to inspire other adults as well as needing to be a good Head Shepherd to a large flock. It would be hard to argue against the case that a comprehensive school provides more opportunities for leadership in its teachers than a grammar school together with one modern school or more. A school of 700 may recognize and use a Deputy Headmaster, a Senior Master and four Housemasters, all men in authority and with great responsibility. A school of 1,350 may depend on 10 Housemasters. The quality of the teaching profession is shown in the way these demands for top qualities in key-posts have been met to the full.

Of course, the Housemaster has to delegate authority to the tutor in charge of the group of 30. The Housemaster trains the tutors and supervises them, keeps them on their toes and pats them on the back when deserved. Complaints from parents come first to the Housemaster; complaints about the individual boy may mean writing to the parents, or asking them to come for a talk, or going to see them at home. He is responsible for getting essential information through the tutor to the pupil and information at staff-level to the tutor. House assembly, House dining are his responsibility. A precious blend of efficiency and humanity is called for, a balance between initiative and loyalty to his Headmaster, a combination of strong hand and velvet glove. He has to take the right tone with the youngest staff and with the oldest boys. He can be busy on House matters throughout the day, with dinner-time no breathing space, and come back in the evening to be free to receive parents during 'Surgery', the fixed hour or so when he attends to deal with aches and pains. When he leaves, there is a gap: someone else must begin to get to know all the boys and almost all the parents.

Variations of the House pattern are possible. There can be a Lower School Housemaster, caring for the first- and second-year pupils in separate buildings. One school with eight Houses recognizes a Deputy Housemaster in each House. In that school the declared aim is that every child shall be known personally by at least six members of the House staff. Another school recognizes an Assistant Housemaster in each of 10 Houses. Some mixed schools appoint a woman as well as a man to an official position at the head of the House. A different school combines Year-master with Housemaster, the latter relying on the former for settling questions of re-setting.

The sixth form gets separate treatment in a few schools, particularly where it is possible to offer the sixth their own building or section of a building. In one school the sixth pass from the care of the Upper School Housemaster to that of a sixth-form Master and have their separate assembly.

That enables fifth-form boys to take the senior positions of responsibility in the House. In another school the sixth is regarded as an additional and distinct House.

Objections have been raised to having too much contact between the small boys of 11 and the tall young men of 18. Isolating the sixth is one way of meeting such objections. The other way is to isolate the small boys. For example,

New entrants spend their first two years in one of the Lower School Houses, amongst boys of their own age. They are then divided among the six Upper School Houses, approximately 50 to each House. It can be argued that the break and change of House at the end of the second year could cause a great upset in a boy's school life, and that 10 vertical Houses, each containing boys of all age groups, would provide a better pattern. Experience has shown that keeping the younger boys together for two years enables them to settle into the school more readily and more harmoniously and that the break very rarely upsets a boy.

Tutor and group

In an increasing number of schools the tutor-group is the constant factor in a child's school life. The new boy joins a tutor-group on his first day and remains in it. The group is the result of careful selection to give a cross-section of the community; it is part of its essence that it is a mixed-ability group and that the boy remains in it. In this way boys of differing ability come to know each other and to understand each other as they grow up together, however much they separate for subject-setting and when courses lead to different choices of subject. It is also part of the aim that the tutor should stay as long as possible with the group. Separate buildings and staff changes can prevent the aim being reached. Some schools are content with a run of two years for each tutor; other schools work with one tutor for Junior years and another tutor for Middle School years. Those schools giving truly separate treatment to the sixth make a break in tutor-continuity when a student enters the sixth. But the majority work on the principle that a tutor follows his group up the school.

What should a tutor mean to his group? One answer is 'Guide, counsellor, and friend'. Another is 'Provide the leadership, example, and care ideally exercised by the head of a family group'. The tutor must be the link between the staff of the school and each boy in his group. He must grow into the position of being the adult to whom the pupil first turns when in trouble; he must win trust and confidence. Equally important is to be a reliable and efficient channel of communication. It may be better for the individual boy to receive official information not from the platform of a big assembly hall but in the friendly atmosphere of a family gathering, where he can ask questions and a friend can answer them. The tutor too must collect and store all necessary information about the boy so that he can deal with enquiries from other members of the staff and can give good advice when decisions have to be made about course or career. The tutor represents the school leg of the triangle

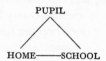

The tutor makes sure that each of his group feels that one adult is deeply interested in his progress in school work.

That is the way to put it—in terms of pastoral care. It is possible to squeeze out the ideals and present the duties as a collection of chores:

(1) Mark the register at the beginning of morning school and of afternoon school. Notify office of absentees. Collect absence notes.

(2) Keep record-cards accurate and up-to-date.

(3) Supervise a section of the dining-room.

(4) Prepare reports for the Housemaster to sign.

(5) Check and initial homework diary.

(6) Arrange rotas for monitors, etc.

(7) Give out notices.

(8) See that school uniform is worn.

(9) Put over propaganda for school societies, school shows.

It would be very wrong to present this as a set of trifles which get in the way of the job of teaching. Tutoring is at least as important as teaching a subject. It should be clear that a tutor begins by doing all that a good Form-master does and goes on to do a lot more. He can hardly dodge any of his opportunities without feeling that he is letting down those who put their trust in him. Because he can direct his energies into so many channels, he deserves to have more free periods than a man who teaches a subject and does little else.

Some schools wisely make time for tutorial meetings. One way is to take the minutes allowed on the time-table for morning assembly and see that on one day of the week the tutor group has its own assembly in the tutorial room (with many minutes gained because there is no movement). Thus the time-table for assemblies can be like this

	In Hall	In Houses	In Tutorials
M.	I, II	—	III–VI
Tu.	Houses A, B, C	D, E, F	—
W.	Forms III and IV	—	I, II, V, VI
Th.	Houses D, E, F	A, B, C	—
F.	Forms V, VI	—	I–IV

One school allows 35 minutes for tutor-groups on the one day when there is no other assembly. Another school provides one period per fortnight for tutorials. The good tutor does not think of asking 'What shall I do to occupy the time?' In a school which bases its pastoral care on the tutor-group, a man's readiness to learn to be a good tutor is considered at interview as important as his ability to teach a subject.

A tutor's life has been described as 'continuous opportunity for forging personal relationships'. One tutor's experience is 'An intimate bond between child and tutor is built up over the years and under this system a vital link between child, home, and school is established'. Another tutor writes, 'One can develop personal relationship on a far more friendly basis than with a class one teaches. Discipline is both freer and easier to maintain.' If the feeling a visitor has on entering a comprehensive school is that it is a happy school where children are friendly and well-behaved, most likely he has come to a school with conscientious, devoted tutors.

Rewards and punishments

The word 'discipline' crept into the previous paragraph. The comprehensive school recognizes that school discipline is not an emanation from the top, falling like rain from heaven alike on worthy and unworthy members of the teaching staff. It grows upwards, rather, from the tutor-group level. If little devils crop up at the rate of one in fifty (or any figure you prefer) a school of 1,900 will have more of them. Moreover, the large numbers can make it easier for the trouble-maker to be undetected. Even when he is detected, it may be by a member of staff who does not teach him, does not know his name. So the larger schools set up machinery to ensure that both commendable and discreditable effort or work is recorded and earns its just reward.

System is essential. If the school issues a termly report, there must be means of collecting and recording essential

49

information along with subject teachers' comments. If an employer asks about a boy, if the boy himself asks for a testimonial, an up-to-date record to which reference can be made is needed. The tutor as representative of Housemaster or Year-master or Head of School is the man who keeps the record up-to-date. Some system is needed to ensure that evidence of good or bad action reaches the tutor.

In the Lower School of one school, 'Stars' are awarded for outstanding progress, initiative or merit in work or conduct. That illustrates the pattern. The boy who receives the 'star' knows that his efforts are recognized, and his friends know it too. Such semi-public recognition is a form of care for the individual because it is a form of encouragement. Before the tutor can know and record the award of a 'star' there must be paper-work. At the tutor's end there must be a card-index system or a file. At the staff end, there must be some standard form of message-card.

Coloured cards are used in one school:

The teaching master will fill in a Pink Card when a boy has done good work for that boy's ability. He then sends it to the Head of Department who passes it on to the Headmaster, who passes it on to the Deputy Headmaster, who sends it to the Housemaster, who gives it to the boy to take home. After three or four of these, the Head asks to see the boy's exercise books and perhaps will give him a Headmaster's Commendation on printed decorated paper in assembly. The white card is for consistently bad work. It goes through the same process but is not handed to the boy.

In another school:

Every teacher has a pad of small printed cards on which to fill in a child's House and tutor-group. If he has occasion to praise or admonish a child, he makes a brief note on the pad and the note is passed to the tutor. The tutor then congratulates the child or discusses the offence. Usually this is sufficient.

A tutor can take the matter further—to the Housemaster or to the Deputy Head or to the Headmaster. The better the relationship between boy and tutor, the less likelihood of any need to take the matter further.

Where the House system is strong, the recognition of good and bad is built into the competitive system. The individual's receipt of a Merit or Demerit or 'Merit for good work or good social behaviour, Penalty for bad work or anti-social behaviour', counts for or against the House. There may be a House trophy or a Work shield for the best House. Blacks (Demerits or Penalties) may be deducted from the score of Whites to decide which is the best House.

One school tries to determine which House is best in effort. A Diligence Review is made every three weeks. In each subject each boy gets a mark for effort made during the period. 'Marks on scale 1, +, 2, −, 3 score 1, $+\frac{1}{2}$, 0, $-\frac{1}{2}$, − 1 respectively.' This system 'encourages boys to try harder'.

One school seeks to encourage good habits by distinguishing some minor offences not as bad conduct but as bad routine. 'If a pupil is careless over bringing books or equipment to lessons and does not respond to admonition, he should be given a Routine Mark.' If 'oral correction' is not sufficient to end a bad habit, a Conduct Mark (regarded as more serious than a Routine Mark) is given.

The same school works a *Notch* system for prefects carrying out their duties. Lists of boys in tutor-groups are ready to receive notches. 'Every case of misdemeanour which, in the opinion of a prefect, demands punishment will be noted briefly, dated and initialled on the lists opposite the name of the child concerned.' Once a week prefects take to the tutors concerned lists showing such black marks or notches. The tutor takes appropriate action: more than one notch may mean lines or detention or taking the offender to the Housemaster. This is an attempt to secure uniformity among prefects in maintaining discipline.

When improvement does not follow initial treatment on

the lines already discussed, an offender can be put 'on report'. The boy takes round with him for a stated period a report card which he gives to the teacher at the beginning of a lesson and collects from him at the end. The teacher may be required to do no more than add 'Satisfactory' or 'Unsatisfactory' at the end of a lesson. One school has separate Conduct Cards and Work Cards for boys 'on report'.

Behind these ways of dealing with those who give trouble stands the detention system, which is little different from what is found in other kinds of school. It is exceptional for prefects to have power to put in detention, except where there is House detention for misconduct and the House is responsible for behaviour in a certain area. A few schools stipulate personal or private detention for bad conduct in class or very bad work in class, leaving school detention for offences outside the classroom. Very much is done to deal with trouble at an early stage, to stop it developing to a point where an official detention is necessary.

Caning is the very last resort, to be used only if every other approach fails. The existence of the deterrent seems to be strong enough in most schools so that there is little demand. The Headmaster usually delegates authority to cane only to his Deputy, the Heads of House, the Year-masters and very occasionally to Heads of Departments. Younger schools are trying to avoid making caning part of tradition. One such school reports: 'There is no cane in the school. The main form of punishment is a work-party. It is a sign of the happy atmosphere and good relations between staff and children that this work-party is seldom needed.'

Careers

Recognition is growing that careers guidance is a special kind of pastoral care. It is aimed at helping the student to choose sensibly and realistically. It entails having all the necessary information about the individual boy and having all the necessary information about the career in the mind of the individual boy. So all the information that has grown as the tutor moves up with his group is raw material for the care of the individual. So some members of staff train to be experts in knowledge of requirements and procedure.

A small school may be able to leave most of the work to one man. More schools now have not only a Careers Master but also one or more assistants. Some schools put one man in charge of fourth-form leavers, another man in charge of fifth-form leavers. A sixth-form Master sometimes deals with those in his care. Mixed schools spread the load by bringing in the mistresses to advise girls. One school asks the Form-teacher to deal with straightforward cases, bringing in the Careers Master when necessary.

Standard features of careers guidance include:

(a) Preliminary interview with the individual boy at the beginning of the school year in which he intends to leave.

(b) Discussion with the members of staff who know the boy very well—tutor and Housemaster or Year-master or Head of School.

(c) Provision of a place where the boy can go to see what printed matter will help him. Notice-boards and display-boards to show what is more recent than what is in book form. Provision of material which can be borrowed to show parents.

(d) Full use of official services of the Youth Employment Service; inviting their representatives to give talks, arranging personal interviews, making the services available to parents at Careers Conventions.

(*e*) Liaison with local Technical Colleges and local employers.

It is in the schools most influenced by the Newsom Report that the most is done. Here the programme of studies has a deliberate weighting of lessons looking ahead to preparation for beginning work. Careers truly comes into the time-table. Here visits in school time are one form of care of the individual, in bridging school and world beyond school. The works experience of some schools is another example of the thoroughness of this pastoral care (see chapter 8, p. 95).

6

THE FOUNDATION YEARS

THE FIRST YEAR

Every comprehensive school tries hard to begin well—to organize the teaching so that each boy or girl on reaching the fourth or fifth year is working well on the right course and at the right pace. At the present time the first year in comprehensive schools is a testing ground for the virtues of streaming and setting in comparison with mixed-ability groups or no streaming. At present, each camp is confident of being right. Final decisions must wait until the present unstreamed first-year pupils have worked through to the end of the course.

There is often division of opinion inside one school. Sometimes it seems that if a vote were taken among those who take G.C.E. work the majority would be in favour of streams or sets; if the vote were taken among all the staff there would be more in favour of mixed-ability groups. Representative opinions of our members include:

(a) 'The children are in fact stratified within their own sphere of friends.'

(b) 'Many Heads of Departments are pressing for non-streaming and a full academic syllabus for all children in the first two or three years, so that all have an equal chance of academic work later. We do not accept that there is ever any "levelling-down" by non-streaming but that a sense of failure is minimized and ability in any direction is capitalized.'

(c) Non-streaming is 'well suited to all boys below the fast-fliers or grammar school "A" stream'.

(*d*) 'Streaming possibly benefits the abler pupil.'

(*e*) 'No noticeable feeling between pupils of high ability and pupils of low ability.'

(*f*) With streaming 'there is positive incentive for promotion within the school'.

(*g*) 'Streaming has not had the most beneficial results. It has created "sinks" and they have contributed to bad discipline.'

(*h*) 'The *possibility* of promotion has its effect. The side-effects are purpose, incentive, fulfilment or snobbery, disillusion, apathy.'

The active House system combines well with mixed-ability groups. In one such school the L.E.A.'s I.Q. rating is used to determine the most backward (21 out of 320). They become one set in one House. The rest are divided into two parallel lists of 150 each (for two Houses). Each 150 is then divided into five House groups (each under a Group Master) so that each set of 30 is a cross-section of the whole 300. The group stays as a group for Arts and Crafts, P.E., Games and Swimming, Physiology, Drama, Lecturettes. Each 150 is separately divided into five English sets as determined by standard English tests. The time-table allows each House to be taught Maths at the same time and for the top three English sets of each House to take French at the same time. Sets in French and Maths are formed during the first term. Sets four and five of each House have General Subjects under one teacher, not separate English, History and Geography.

Remedial classes

Separating the most backward pupils is a key feature in almost all large schools, even where the rest are unstreamed. The name may vary, but the principle is the same. The aim is to bring the special cases back into the main stream as soon as possible. One school (12-form entry) has three Remedial Classes in the first year, two in the second year, one in the third. The numbers in Remedial Classes may be limited according to the number of trained teachers in the school, for the pupils need 'specialist' teaching just as they need separate specially equipped rooms and books which are different from those used by other pupils of the same age. They need more practical lessons—in smaller groups. Keeping the numbers small (under 18 in one school) may cause other classes to be larger.

In one school the whole idea of separation is being challenged.

We began last year with a full-blown Remedial Department, with its own classes and specialist rooms. Many members of staff felt uneasy about this situation, becoming gradually more and more convinced that the attitudes it set up in these pupils were working against what we were trying to achieve. This year we have no 'Opportunity Classes' as such in the first year. Instead, the remedial teachers work closely with the teachers of English and Maths giving help either in the classroom without removing the pupils or in prep. groups outside the Maths and English rooms. This means that because there are no 'Opportunity Children', segregated from their fellows and unable to do French and Science, there are fewer behaviour problems and less difficulty in maintaining a basic course for all. Previously a child sent forth from the Remedial Classes was a problem simply in that he was so far behind in subjects he had missed or had never started.

Initial selection

Pupils for Remedial Classes have to be selected on the results of some kind of testing or rating. To ensure a cross-section in a mixed-ability group, there must be some evidence of measurement. Only the few schools that put all their first-year pupils into batches according to alphabetical order can dispense with some assessment. Five English schools rely on their own objective tests. One school arranges for the September pupils to spend a day in the previous summer term doing tests to provide a basis for streaming. Fourteen schools rely entirely on primary school reports or recommendations. Sixteen schools combine primary school material with objective tests. Two schools supplement written material from primary schools with interviews with the primary school Headmasters. One school draws up provisional lists on the evidence of L.E.A. tests and invites Headmasters of supplying schools to comment on the distribution. One school combines primary school recommendations with verbal reasoning tests. In one school an interview is arranged to bring together the Senior Housemaster, the primary school Headmaster and the parent—15 minutes for each interview. One school allows a settling-in period and before the end of the first month sets tests.

Partial selection

A number of schools avoid the word 'stream' and speak of 'bands' or 'groups'. For example,

Children entering the school are placed in one of three groups: (1) selective pupils; (2) non-selective pupils; (3) remedial pupils. The selective pupils are divided arbitrarily into four or five forms and follow a full academic syllabus for three years. French and German are taught in the first year: Latin is introduced in the second year.

The non-selective pupils follow a less academic syllabus, with no foreign languages and more practical work. In theory they may be promoted to the academic stream but in practice the number is small.

In very few instances would the descriptions 'grammar school forms (or streams)', 'modern school forms (or streams)' be used but in those schools where organization is in bands there is a division between upper and middle. 'Academic' is not a popular description for this upper region, the basis of which is almost always the full range of subjects taken at 'O' Level in G.C.E., including a modern language but possibly excluding Latin. When we read of parallel forms within the band, or mixed ability within the band, we should not forget it is unusual for a boy to move from band to band.

Many schools are ready to try to improve on their ways of organizing the foundation years (one school offers a stimulating 'motto'—'We do nothing here because it has been done here in the past'). A picture of a school in the process of developing its 'band' system is seen in the following:

We used to stream into six forms as a result of junior school reports and our own tests. In 1965 for the first time we have made three bands. Each band consists of two forms of equal ability. (Alphabetical division within the band.)

In the first three years the subjects taught are the same for all forms (with different material and presentation according to the ability of the pupils) with the following exceptions: (i) French is introduced to the top four forms in the first year. (ii) German is introduced as an additional language to the top two forms in the second year. (At one stage we used to give Latin to the top stream and German to the second stream. We have abandoned this and Latin will be introduced in the Third or Fourth Year for those who want it or seem likely to need it.)

One school plans in terms of four bands:

>A (4 forms)   B (1 form)   C (3 forms)
>D (Remedial, 1 form).

Other three-band divisions are:

(i) Above average	Average	Below average
(ii) Grammar or academic	Non-selective	Remedial

Concealment of grading

Whether division is in descending order of ability or in bands, each form has to have a description or a name, bringing up the problem of pupils themselves appreciating their rating. In answer to a question about trying to conceal from pupils the implication of ability-distinction, one correspondent wrote 'What would be the point? They all know within twenty-four hours.' Several schools said that pupils were not told, neither was the situation concealed. Some have faith in a code—for example:

Used	C	O	N	S	T	A	B	L	E
Translated	A	B	C	D	E	F	G	H	I

Though the staff, like workers in a shop with private marks on goods, soon get used to the system, parents guess the answers (but not necessarily the correct ones). The naming or numbering of sets involves the same difficulties:

The changes in nomenclature of the sets each year often very effectively conceal the ability of a set from the teacher until he has had some experience of the set in question. However, children are very astute over this sort of thing and they catch on very quickly, so that the terminology adopted for a year is obsolete by about February or March and is replaced for the following September.

One wonders if a set of codes has to accompany the school's record cards or filing system for showing the progress of a pupil. Another school believes that it is possible to label the forms or sets so that there is 'no obvious or immediate appreciation of ability', still another school insists that knowledge of grading does not cause friction because 'pupils know that no matter what form they are in they stand an equal chance of getting a good teacher'. Where it is part of the basic approach that '"less able" does not mean "less important"', where there is nothing equivalent to 'streaming the teachers', there will be greater acceptance by pupils of their place in their year. More than one teacher says realistically of pupils, 'They have sufficient intelligence not to be deceived by subterfuges'.

It is exceptional to find a fast form, an express stream, in a comprehensive school. There is strong objection on principle to making a special class, so markedly different in programme and pace, so unmistakably taken by a special kind of teacher. Only two schools arrange the work of the first year with a view to deciding which pupils can take G.C.E. 'O' levels in four years.

Setting

Only a minority of schools in England have unstreamed, mixed-ability groups in all subjects. Thus there are many variations in the degree of setting in first-year work. For example:

Setting is developed gradually throughout the school. There are eight mixed-ability first forms in four Houses. After a time, which varies in different subjects, sets are arranged within the Houses (Set 1, Set 2, in each House). Maths and English are set across four forms.

English and Maths are the first subjects to be set in many schools:

On entry, all pupils are given our own tests in English and Maths, and placed in eight sets by ability, called like the groups A1–A8. (Thus Group A1, English A1, and Maths A1 could be three quite different formations, although a number of individuals will be common to all three.) Pupils will thus work in a homogeneous group of comparable ability, and can be moved up and down as and when performance suggests. Such movement is kept to a minimum to ensure stability, but the aim is flexibility in the system. Subjects other than English and Maths are taken in English sets, but where staffing permits two or more sets to be time-tabled together, the Head of Department is at liberty to make changes within that block of sets. Always flexibility is the aim, so that each pupil can be catered for at the appropriate level...Pupils are fully aware of the implications of setting but provided we have setted correctly no grievances occur.

Other ways of introducing setting include:

(i) A common course for three years, streamed according to ability in English but sets for Maths.

(ii) Sets for English and Maths; for the rest of the time in broad horizontal bands of ability.

(iii) Unstreamed for most subjects but set for French and History.

(iv) Setting for English determines the sets for History, Geography, Modern Languages; setting for Maths determines the sets for Science.

In a few schools setting results in smaller classes. In a small school, the bottom two forms become three sets for English, so that classes are 15–20. In another school, three streams are split into four in basic subjects (English, Maths, Science, Geography, History)—'It appears to allow each pupil to work within his ability in each subject'. In a third school the top two groups become three sets for English, Maths, Science and French.

Setting has its advocates and opponents. In one school, we are told, 'most of the staff favour setting for academic work'. Another school has cut down setting in the first two years, 'where it was felt that it led to a form losing its "team spirit"'. In a third school, 'Most of the staff agree that too rigid a system of streaming and setting is only reproducing a bilateral system within one building. They consider this undesirable and are making efforts to break the problem down.'

Common course

Attempts to do two things together create difficulties. Some teachers wish strongly to have homogeneous groups for teaching—that almost always mean 'academic subjects' to use the term which slips in now and again. Other teachers wish strongly that a pupil with a high I.Q. should mix from time to time with pupils with a lower I.Q. One school reports that in the first year one-third of the time is spent in mixed-ability groups: probably covered by Art, Music, Physical Education, Woodwork or Domestic Science, Religious Instruction. Our members know less about the practical side than about the classroom side in what are termed 'basic subjects'. Their eyes are kept on the subjects which are the staple diet of a full 'O' Level course in G.C.E. There is a commendable desire to arrange a time-table for the first year which will not limit the number of subjects a late-developer can take at 'O' Level. Every possible provision seems to be made for a boy who starts in the middle stream (when streaming operates) to be able to climb up if he deserves to. Consequently in the first year there is readiness to start all pupils above Remedial Class in subjects from which some will properly retire at a fairly early point.

So a Common Course of subjects is frequently referred to under different names. In the first year it can mean those subjects which will be in the 'O' Level course or the C.S.E. course. Where classes are unstreamed, where the particularly able boy can be in any subdivision, the programme in terms of subjects must be the same. Where 'bands' operate, the subjects can be much the same with differences 'in breadth, not in depth'. The term 'General Education Course' is used in very few schools to cover Music, Art, Drama, Physical Education, Handicrafts, Religious Instruction; in most schools it means all the other subjects. A school compelled to separate its Junior School for three years in buildings one mile away from the rest of the school is in a position where a Common Course is needed for three years. Usually two years is the duration. The disadvantage of these descriptive phrases is that the language problem is concealed.

Languages

One school is able to co-operate with the primary schools that supply it and arrange for French to be started before the pupils come to it. All other comprehensive schools in our survey make a beginning in French in the first year. Seven schools report that the whole of the first year take French; three Schools specify 'all but the least able' (or E.S.N. pupils or the bottom group). Ways in which French becomes one of the languages started include:

(i) Five forms French, three forms German.

(ii) Two forms French, seven forms German (out of 10).

(iii) Three forms French, four forms Spanish.

(iv) One of French, German, Spanish done by each form.

In all except (iv) it is the top forms or groups that take French. Spanish or German shares with French in two schools but the distribution is not stated. In one school four of six groups take French, in another seven of nine, in a third two of four groups. A different school begins with four of six groups taking French and intends to keep it like that for four or five years. Several are ready to reduce the number of forms taking French in the second year. One school brings in Russian in the second year. Though it is too early to report on the full effect of C.S.E. on work in comprehensive schools, we expect the ability to take French in an external examination at a level lower than that of the

G.C.E. (and an examination in which the school can go a long way towards determining the syllabus and the assessment) to encourage the retention of French in the middle groups. There are signs of efforts to make it rare for a late-developer to be unable to reap the full benefits of promotion because he had 'not started French'. If French is the weakest subject with a late-developer, it is likely that there will be at least a C.S.E. directed group which he can join.

Latin is becoming more and more a subject found mainly in comprehensive schools which absorbed a grammar school. Three schools in our survey introduce Latin in the first year; four schools introduce it in the second year, while in a fourth school the addition is Latin or German. One school brings in Latin in the second term of the first year for a top stream and in the first term of the second year for the next two streams. Where the school does not stream or set its pupils, Latin is unlikely to be taught.

Homework

'Homework is compulsory for all except Remedial Classes' is representative of the position in comprehensive schools. With non-streaming, any of the forms in which a particularly able boy can be found—that is all above Remedial—must have homework, apparently. The load is light, almost nominal in the lowest groups based on ability rating. Homework need not mean written work—much is left to the discretion of the individual teacher.

Some schools take homework more seriously. Some pupils are required to keep a homework diary and to enter the work set for each night. Such a system is incomplete without arrangements for regular supervision by tutor or Form-master or the parents. Attitude to homework is one item which may be considered at the end of the first year when arrangements are being drawn up for streams or sets in the second year. In the first year, one aim is to lead the pupils to accept homework as a standard feature of school life. Fifteen or twenty minutes per subject (as in some schools) is an almost painless introduction.

Redistribution

At the end of the first year, it is hoped, the dispositions and capabilities of individual pupils will be known and the second year will begin with a stable position in a group for almost everyone not in a Remedial Class. One way of placing pupils is to regard the initial distribution as a rough sorting out and to be ready to make changes whenever the justification is established. The more rigid the streaming, the greater the need for very full consideration of each case. With setting, along with simultaneous timetabling of one subject, a Head of Department may be allowed to make changes after consultation with the teachers concerned. The other way is to take the greatest care with allocation after December. Then the children have had time to settle, the teachers have had time to study their ways and estimate their potentialities; a full staff meeting should be able to arrive at just and acceptable decisions. One school with an intake of 360 needed to move only one boy after its decisions at the end of the autumn term (1964).

Teachers are clear about the purpose of redistribution and careful about the terms used. The purpose is not to stress differences but to place each boy or girl where he or she can work at an appropriate pace in a suitable course. One school insists that 'promote' and 'demote' should be avoided; what they do is 'regrade'.

Different ways of checking initial decisions include:

(i) A staff meeting at the end of October to consider obvious misfits.

(ii) Staff meetings to consider changes (*a*) at the end of the first month, (*b*) at the end of the first term, (*c*) at the end of the first year.

(iii) Full review of first-year pupils twice a year.

(iv) Gradings made every half-term. Changes made if these furnish evidence of need for change.

Two points are worth noting. One of the few schools with a 'fast form' deliberately begins under-strength, with a small number, to allow additions (regarded as 'promotions') after December and April. Mixed forms have problems of balance between boys and girls. Sometimes a change can be made only if there is room in the receiving form, for no class can have more than 20 boys or 20 girls if the half-class is one unit for work in a Woodwork room or a Domestic Science room which cannot accommodate more than 20.

The advocates of non-streaming think that pupils are happier and suffer less from anxiety and strain when they do not have to wonder what has been done at 'biennial re-setting sessions'. Moreover, the staff in schools with the minimum of grading give up much less time to consultations, preparation of figures, and formal meetings.

SECOND YEAR

With the second year, increasing provision can be made for the differing needs of pupils. There is somewhat more justification for the use of the terms (only in a few schools) 'Grammar or Academic' for the top groups. One school describes its second-year divisions thus:

> 2 streams—G.C.E. directed.
> 2 streams—C.S.E. directed.
> 2 streams—probably below C.S.E.
> except in certain subjects.

Another school with ten groups arranges them with two sets to each of these streams:

(A)   Potential G.C.E. standard.

(B)⎫
(C)⎭   Potential C.S.E. standard.

(D)⎫
(E)⎭   Non-examination.

A different school retains its common course for both second and third years, with no choice of subjects, because 'We try to delay specialization until children make the correct choice and do not need to change'.

An example of how some subjects can have different numbers of periods is shown below, in the programme for the second year in a small country school. For registration, etc., there are five mixed-ability groups, one for each of five Houses, retaining the same Form-teacher for two years. For teaching purposes they are divided into 'ability groups' (with sets in French and Maths):

	English	History	Geog.	French	Latin
2 A	5	3	3	6	4
2 B	6	4	4	6	—
2 C	6	4	4	6	—
2 D	6	4	4	—	—
2 E	6	4	4	—	—

	Maths	Introd. Science	D. Sci./ Wood	Needle/ Prac. Agri.	Art and Music
2 A	6	4	2	—	4
2 B	6	4	3	—	4
2 C	6	4	3	—	4
2 D	7	4	3	2	4
2 E	7	4	3	2	4

Among other second-year arrangements are:

(i)   Nine forms, the bottom (Remedial), next to the bottom ('Weaker pupils'), above them seven forms unstreamed.

(ii) The pupils graded for English—the English group counting as the form-group. Sets in all other basic subjects.

(iii) After no setting in the first year, 'sets' are begun in English and Maths in the second year.

(iv) Half in parallel forms in bands

> 2L 2M 2N (set for Maths/English)
> 2O 2P 2Q (set for Maths/English)
> 2R 2S 2T 2U 2V 2W (Not set at all.
> Taught in forms for all subjects.)

In the section on Languages we referred to the introduction of a second language in the second year. One school brings in Russian for one form, German for two forms. A few schools offer a choice of Latin or German (to two forms to make one group of each). One school offers options to the top four of eight groups—one of Latin, German, Spanish. In a few schools the very bottom groups drop French after one year's experience. It is unlikely that the boy who will need a foreign language at a later stage for University Entrance Requirements will not have made a start in one language in these foundation years.

7

THE MIDDLE YEARS

What one school says of its work throughout the school applies particularly to the third year and the fourth year in comprehensive schools: 'The prime aim is relevance—that is to estimate the needs of the boys and girls within the immediate or fairly immediate future, then to educate them to meet their needs in as stimulating and interesting a way as possible.' The needs of each individual pupil, this school believes, are (a) in relation to his or her future career, (b) as a future home-maker, (c) as a member of a community, (d) as one who will have leisure-time. Programmes of school activity based on such needs will not forget external examinations but will use such examinations as a means of satisfying needs.

These middle years become years of careful preparation for different kinds of children. Differences in aims and needs of these different kinds become more marked and more important. Each school works out the best ways of fitting its children for the future. As the balance between different kinds of children varies from school to school, what succeeds in one school may not succeed in another. There is less of a general pattern throughout schools in these middle years and more illustrations of tailored outfits, single-school attempts to do justice to all—the bookish child, the child gifted in practical skills, the problem child uncertain of aim and with abilities flattered by being called 'moderate'.

In some schools the onlooker is struck by the care taken with the bookish child, to repeat a phrase chosen to avoid arousing those feelings that 'grammar-school type' or 'academic' liberate when used of normal boys and girls

70

who enjoy and do well in classroom lessons that depend much on reading in books, learning from books, writing in exercise-books. Some of our members think that such children do much better in comprehensive schools because they are prevented from spending *too much* time in book-based lessons. Art and Music, Woodwork and Domestic Science remain and are often compulsory. This is accepted in an atmosphere where there is no suggestion that they are minor or subsidiary or peripheral. For example, in a school which offers Latin or German in the third year, two periods are still provided for Technical Drawing.

The programme in the third year for top groups is more clearly looking forward to G.C.E. work at 'O' Level and to building up a sixth form. One school now starts two of its twelve streams as express streams. Those pupils who have shown proficiency in one language now have a chance to start another—Spanish or German or Russian or Italian at one school, according to what the staff can supply; a second language is studied instead of double science at another school; in some schools pupils who began with Spanish can now add French; streams 3–8 of a 12-stream school now start a second language. The option system which offers a language to some but not to the whole group can mean that choosing a language prevents such pupils from doing Biology or Chemistry or Handicrafts.

Science

Teachers in grammar schools, interested in the teaching of Science in comprehensive schools, ask 'Are the separate Science subjects taken to "O" Level?' The answer is 'Yes'. If the askers go away satisfied they are short-sighted. The supplementary question, 'By whom?' is more important. 'Science' as the description of what is learned means different things at different levels. Teachers in comprehensive schools believe that Science is an essential item in the diet and take pains to provide it—at the right level. For half the pupils in the third year, lessons in Science preparing for 'O' Level in separate Science subjects might be quite wrong. Third-year programmes cannot leave out of reckoning the certainty that two out of five pupils starting the third year will leave at the end of the fourth year. (see Table 4 on p. 82). Science presents the problem in a clear form—work has to be planned in any subject to meet the needs of:

(1) Those who have only two years more of school.

(2) Those who have three years more of school.

(3) Those who have decided already to spend three years leading to 'O' Levels and to continue at school in the sixth form.

The section most difficult to cater for is the middle one because it includes:

(1) Those who can try a few subjects in C.S.E.

(2) Those with hope of success in a full range of C.S.E. subjects with one or more at 'O' Level in G.C.E.

(3) Those who will have more subjects at 'O' Level than at C.S.E. Level.

(4) Those who will take a full range of subjects at 'O' Level—as if they were in a grammar school.

In the first two years—particularly where mixed-ability groups, not sets or streams, are arranged—there is a course in Science which can be taken by all and which will provide a good foundation for all the developments which come in the third year in most schools, in the fourth year in the rest. In every year what is provided is suitable Science. What is called 'General Science'—not the same as preparing for General Science at 'O' Level—is arranged in many schools for leavers, for 'weaker groups', or even for the lower half of the whole year. The schools providing General Science for all for two years almost equal in number those extending the general course for three years. So separate subjects begin for suitable pupils either in the third year or in the fourth. Three separate subjects are rarely available. Some schools offer a choice between Physics and Biology, just as some pair Chemistry with Music or Art. Country schools provide Rural Science both for C.S.E. and for general courses. A few schools work towards Physics-with-Chemistry as one 'O' Level subject. The position of Biology seems to depend on the qualifications and interests of the staff. An exceptional school provides Biology for all in the third year, with Physics *or* Chemistry for top groups, with General Science for lower groups.

How the Science teaching is arranged in a small school (5-form entry) is shown below:

(1) The lowest group (no French) has five periods General Science.

(2) The next to the lowest is taught as a form—no setting —and has two periods each for Physics, Chemistry, Biology.

(3) The top three groups have two periods each for Physics, Chemistry, Biology, in sets.

Grading and courses

In the third year some schools plan in terms of subjects; in many more schools the planning is in terms of courses. In the subject-based plan the mixed-ability grouping comes to be limited to arrangement in forms for registration, while teaching is in groups graded in ability. Some Heads of Departments think of 'setting as the normal thing in the third year to cater for C.S.E. and G.C.E. courses'. In large schools there is retention of the 'band' system with parallel forms in an 'O' Level band and parallel forms in a C.S.E. band.

One school separates those pupils who choose a Secretarial Course—the only named course in the year. Otherwise, 'Pupils choose subjects rather than courses, because we feel this gives a freer choice. Both parents and pupils are fully consulted beforehand and career intentions are always taken into account.' One school requires those intending to turn to a Commercial or Secretarial Course to pass an aptitude test for shorthand and typing. Behind this is a strong feeling that whereas transfer from set to set is still easy in the third and fourth years, 'transfer from one *course* to another is a matter of gravity seldom contemplated and only after consultation with the parents and the Housemaster'.

Where everyone in the third year joins one course or another, the pupil makes a choice of one course from a range of three or one course out of four. An example of three courses is:

(A) Three groups taking a common G.C.E. course along with a language or Art, Craft or Music.

(B) Five groups, in sets. 'C.S.E. or equivalent.'

(C) The group that becomes, in the fourth year, the Newsom group.

A school with four courses uses descriptive words:

Academic   Technical   Commerce   Modern.

The course offered to the four sets of the Academic Group includes English, Maths, History, Geography, French, two Science subjects. Because of the belief in 'a good broad general education' the list does not close there: each pupil must take Technical Drawing and Metalwork or Woodwork (for boys) or Cookery and Needlework (for girls) or two of Scripture Knowledge, Music, Art and Spanish.

Preparation for the fourth year

Easter of the third year is not too soon to start preparing for the fourth year. Where pupils are asked to make a choice, the full range of what is offered must be set down clearly for parents as well as pupils to study. Before a decision is taken about staying on or leaving, parents must receive information about the courses of the fifth year. About 40 % of those who come into the fourth year in September will be beginning their last year at school. Such pupils will be happier at school—contented children cause little trouble—if they can choose some of their activities. The size of a large comprehensive school is an advantage here because it leads to a wider range of useful and attractive activities. So a large school can report of this important year, 'Almost every child does what he has asked for and drops what he dislikes' (that is, unless it is Maths or English that is disliked). Another school has found that 'all sensible courses are possible by doubling'. If a 'slightly vocational-based course' is the aim, the career intended by each pupil about to leave must be known. During the summer term of the third-year information sheets with arrangements for recording choices are issued and collected and worked on to prepare for the fourth year.

A representative letter to parents might begin:

In September your child has the opportunity to start on one of the special two-year courses which we are arranging to meet the needs of children of all abilities. Some will take the 'O' Level G.C.E. Examination; some will take the Certificate of Secondary Education (C.S.E.) Examination; for some children the best course is to take some subjects in the G.C.E., some in the C.S.E.

The advantages of qualifications in an external examination would be indicated, together with some emphasis that staying at school for two years is necessary to reap these advantages.

The parents of children who have only one more year at school can be offered 'a course suited to the needs of those who will take no external examination'. Such a course is sometimes called a Special Course (though really every well-planned fourth-year course is a Special Course). Promise is made of 'emphasis on practical and vocational subjects'. The basis is co-operation: the parent expresses intention, informs the school about career decision, records choice where choice is offered; the school decides the details of the course. The next stage is usually a Parents' Meeting with explanation from the platform, questions from the floor, and, if desired, interviews between parent and tutor or teacher.

Sometimes children have to be asked to think again about their choices and requests—in their own interest. An interested teacher can be full of regret when he sees a capable pupil, one likely to do well in an external examination, determined to leave at the end of the fourth year. The staff have to counter both excessive optimism and excessive pessimism. 'We try to strike a balance between pupils' own choices and their own potential.' This is a point where knowledge of the individual through constant pastoral care enables the Housemaster to be a good friend. Where persuasion has to be applied, it should come from one with authority who is able to take a broad, impartial view. The danger otherwise is from subject masters— 'Unfortunately some canvassing goes on and therefore the advice is not always objective'.

COURSES

One school offers these courses:

Course A. To the top three groups:

Basic subjects—English 5, Maths 5, Science 3, Religious Instruction 2, Physical Education and Games 3, Music 1.

(i) The ablest in the top group of the three add History 3, Geography 3, French 5, German or Science 5.

(ii) The next two groups choose one of these options for a Craft Course:

O. 1. 4 periods—Woodwork or Metalwork or Housecraft or Needlework.

O. 2. 4 periods—Science or Art or German (if begun).

O. 3. French 5 or Geography 3 with Commerce 2.

O. 4. 3 periods—Technical Drawing or History.

Course B. A course with a Technical bias offered to boys who ask for it. Usually 2 groups are formed with C.S.E. in mind.

Basic—English 5, Maths 5, R.I. and Music 3, P.E. and Games 3, History 3.

Technical subjects—Technical Drawing 2, Art/Design 2 or 3, Science 4. Options—French 4 or Geography 3, Woodwork or Metalwork 4 or 5.

Course C. Mainly secretarial, for girls, with some subjects at C.S.E. level, and the opportunity of taking some at 'O' Level.

Basic—English 5, R.I. and Music 3, P.E. and Games 3, Mothercraft and Child Development 2, History 3.

Secretarial—Commerce 2, Accounts 3, Typing 5.

Options—4 periods French or Geography; 5 periods Shorthand or Needlework 3, with extra English 2.

Course D. C.S.E. Course with bias towards Commerce and Retail Distribution.

Basic—English 5, R.I. and Music 3, P.E. and Games 3, Speech Training 1, History 2, Geography 2.

Commercial subjects—Typing 5, Retail Distribution 2, Accounts 2.

Social Studies—7 periods covering Homemaking, Mothercraft, Personal Hygiene, Social work.

Optional—3 periods Art or Needlework.

Course E. A course providing good general education including Social Education. One group will take some subjects at C.S.E. level; another group will have no external examination.

Basic—English 5, P.E. and Games 3, R.I. and Music 3, Maths 2, Social Studies 10.

Girls—Mothercraft and Child Development 2, Home-making 4, Woodwork 3.

Boys—Science 2, Cookery 3, Art 3.

Options. For girls, 3 periods Art or Needlework. For Boys, 4 periods Woodwork or Metalwork.

Another school stipulates for all fourth-year pupils English, Maths, R.I., P.E., one Science subject, one Social subject (English Literature, History, Geography, Civics), one practical subject. Each pupil there has six subjects apart from English and Maths. To enable different children to take a subject at appropriate levels subjects have to be grouped for time-table purposes. These groupings are made (O = G.C.E. 'O' Level; C = C.S.E.; N = no external examination):

A	B	C
French O	History O	Latin O
French C	Chemistry O	Geography O
Art O/C	Eng. Lit. O	Geography C
Geography C	R.I. O	R.I. C
Metalwork C	History C	Physics C
Needlework C	Commerce C	Typing C
Art and Craft N	Typing C	Visits
	Social Studies N	

D	E	F
Physics O	German O	Art O
Biology O	Music O	Cookery O
Biology C	Tech. Drawing O	Needlework O
Human Biology C	Tech. Drawing C	Metalwork O
Chemistry C	Commerce O	Woodwork O

D	E	F
Handyman N	Shorthand C	Housecraft C
Commerce N	Biology C	Needlework C
	Typing C	Metalwork C
	General Science N	Woodwork C

Such a list suggests the expertness needed to make full use of staff and specialist rooms to cater for different levels in a wide range of subjects so that each individual pupil can have an appropriate and satisfying course.

Almost all courses for 'O' Level are for two years. In one interesting exception there is earlier examination entry for the top group of ability-graded groups:

End of fourth year: English Literature, Maths, French.

End of first term in fifth year: Physics and Chemistry.

The utmost care is taken both in drawing up what is offered and in guiding pupils to make the right choice, because it is believed to be in the pupils' interest that there should be the absolute minimum of change of course after the early weeks of the fourth year. 'Transfer is not ruled out but hardly ever occurs'—a tribute to careful placing.

One school offers 15 different courses within a framework of Arts, Science, Technical, Commercial, General Courses. The bigger the school, the wider the variety, then the greater chance of there being a suitable course to which the exceptional misfit can transfer. The choice of a course may be a mild form of specialization and most teachers approve the aim of the man who wrote, 'We try to delay specialization until the children make the correct choice and there is no need to change'.

Additional languages are fitted into courses where staff are available, but only with difficulty. Three schools begin a two-year course in German for some pupils. Of two schools that used to do Latin from the lower school upwards, in one Latin is being reduced to a two-year course beginning in the fourth year and the other is starting Latin *after* 'O' levels, i.e. as a sixth-form subject. In five other schools it is an optional subject in the fourth year. One school offers Greek 'if requested'. One school starts a two-year course in Spanish for 'O' Level, for selected pupils.

LEAVERS

A small school with a three-form entry offering three courses can do little but arrange an 'O' Level course, a C.S.E. course, and a non-examination course. There is a danger in accepting too early a pupil's declared intention to leave if that means excluding such a pupil from groups, sets, or options working for an external examination, to which the pupil may wish to go later if he changes his mind. It may be right to reduce the number of lessons for weaker pupils in Maths and Science and in subjects 'involving the grasp of logical sequences between the work of one lesson and that of a following lesson'. It certainly can help the pupil who changes his mind and decides to stay for a fifth year if sufficient formal instruction is maintained in order to keep open the possibility of joining fifth-year courses.

One school stresses the importance of practical work for those in their last year, as much as the equivalent of two days per week. Variety is important, to increase for each individual the chance to do something he had always wanted to do, something he would not be able to do on his own at home. For example, a school can provide opportunities in film-making and boat-building. It can be asking too much of a Newsom child simply to put one practical activity on the time-table for the whole of a school year—much more unreasonable for two years. Therefore 'we have a term on house-decorating, a term on motor mechanics, a term on work in glass-fibre', still leaving room in the following year for photography, another on less conventional wood-work. The variety must depend on the gifts and interests of the staff at the time.

Providing 'stimulating and relevant courses' is a sound way of preventing or diminishing the malaise that can come to boys and girls of lower than average ability. The boy who is not allowed to forget that he is but second-rate may 'accept it in defeatist resignation' or 'resent it in outraged defiance'. It may seem a simple matter to give him many things to do in the hope that he will find he is first-rate in something; in practice it means thorough preparation and planning, patience and perseverance on the part of the teacher.

The book-learning part must not remain the same for the Newsom child as for those of the other half. The sympathetic experimenters who respond to his needs are right in rejecting any 'watered down version' of an 'academic' course suitable for a totally different child. A teacher of the really able learners looks with horror on obliterating subject divisions; the revolutionary approach is nevertheless right. One school creates an integrated course ('This World of Ours') which is a new blend of English, History, and Geography. Its 'Money and Machines' replaces (and transforms) Mathematics. Its 'Meeting Point' develops Religious Instruction into discussion of issues affecting today.

An additional feature of the planning in this school (working on a 50-period 'week') is a 'special' subject using six periods. It is 'special' in that it is new to the pupils, not just continuing something already sampled, and in that it is designed to be a basis for similar work in the fifth year, then to be more directly concerned with vocation. A boy spends one term on each of these courses:

Handyman   Homemaking   Car Mechanics.

A girl spends one term on each of these:

Handywoman   Homemaking   Commerce.

TABLE 4. *Distribution of pupils over different years*

No.	Total (1)	First year (2)	Fourth year (3)	Fifth year (4)	Figures for previous year (September 1964—July 1965) Leavers Fourth (5)	Fifth (6)
			(A) Boys only			
1	800	150	134	134	55	114
2	710	150	156	80	55	73
3	690	148	131	101	53	50
4	1330	240	228	194	67	101
5	1220	240	196	141	90	72
6	1340	287	294	146	158	111
7	1300	307	292	146	118	132
8	1840	360	335	228	126	114
9	1660	390	354	124	120	66
			(B) Mixed small			
10	530	99	106	39	62	26
11	970	184	221	127	104	72
12	810	199	152	74	46	55
13	600	120	116	55	53	23
14	960	199	167	137	60	76
15	600	84	134	69	52	41
16	960	172	151	106	89	58
17	950	195	203	124	80	91
18	910	170	195	707	73	86
			(C) Mixed large			
19	1920	390	428	226	176	169
20	1430	315	255	162	118	78
21	1420	284	337	130	180	100
22	1380	275	307	137	174	76
23	1940	416	438	161	280	138
24	1480	330	341	99	140	21
25	1450	450	304	145	150	25
26	2040	466	455	214	230	55
27	1310	290	308	93	144	51

The 'newness' of boys taking Homemaking and of girls receiving training and practice in the use of tools should appeal to the pupils. The teaching profession should know more about such pioneering work.

As a bridge between this account of what can go on in the middle years and the following section on fifth-year work, we offer a page of figures (see Table 4 opposite).

Notes to Table 4

(1) The years when the September 1965 fourth forms and fifth forms entered the school may have had a smaller or a larger first form. The supply fluctuates.

Some schools gain through removals, so that the figure in col. 3 can be bigger than the figure in col. 1.

(2) It is not to be expected that col. 3 minus col. 5 corresponds to col. 4.

(3) The size of the sixth form cannot be calculated by deducting col. 6 from col. 4.

(4) These figures are not intended as proofs of anything, only as indications of trends.

8

THE YEAR OF EXTERNAL
EXAMINATIONS

Continuous selection is almost completed when a boy in a comprehensive school begins his fifth year. He is in a unit for registration which may be a form or may be a tutor-group. He mixes with pupils with different kinds of ability in general lessons (P.E., R.I.). But he spends almost all his time in ability groups in his chosen subjects or on his chosen course. For any one examination subject he is in a set or in a parallel form in a band or stream. The initial placing, in third year or in fourth, may be regarded in a few schools as preliminary or provisional. Most schools regard him as settled, though the staff are ready to make adjustments if his progress and attitude call for change.

This fifth year is the year of the first external examinations. But which? Much depends on the subject. The more determined teachers are to draw up for C.S.E. the syllabus they want for the pupils they know, the more likely they are to be tempted to show where the G.C.E. syllabus-makers have gone wrong. The home-made syllabus may be a better syllabus (for the one school) than the one provided by an established examining board. The bigger the differences, then the earlier has the choice to be made whether it is C.S.E. or 'O' Level for John Smith in History. The operating principle is that the decision is delayed as long as possible. Where there are no complications of syllabus differences, the decision can be left almost up to the time for making the entries on official sheets. Many schools have a half-way examination in the Spring term and use the results of this internal examination to check or change official entries in the external one. This internal examination, condemned to carry round its neck the label 'mock', will have an importance magnified or diminished according to the atmosphere of the school. It can be presented as a test of progress; it can be turned into an encouragement or its results can be quoted as condemnation. The candidate's fifth year is doubly a year of anxiety if he feels too strongly that his future depends on both internal and external examinations.

In a small school, where the teaching of one subject is in the hands of one or two teachers, decisions based on the results of the internal examination are fairly and informally reached. In a large school when many teachers, differing in standards and experience, produce marks not easy to standardize, the Head of Department can act as arbiter. In one school a full-scale inquest is held; after discussion and decisions at this staff-meeting, each pupil is interviewed separately.

The composition of a fifth form in a comprehensive school must be kept in mind. Table 4, given at the end of the previous chapter, suggests that out of every five of the first year intake three at most are starting the fifth year. It would be wrong to think of the fifth form as being equivalent to the fifth form of a grammar school with the addition of some pupils who might have taken an external examination in the top form of a modern school. The range of the comprehensive school fifth year is unlikely to go quite so high, if our members are correct in claiming that the very top, the top 5 per cent of the ability range, do not reach them at all. The range goes much below the level of the best form in a modern school; it must go as low as it can go in a modern school. Our members speak confidently of the success of a comprehensive school in bringing on and inducing to remain at school those boys and girls who would not have found their way into the 'O' Level form of a modern school.

There will be many pupils whose interests are mainly practical, so that the percentage of the boys taking subjects from the Metalwork, Woodwork, Technical Drawing group will be larger than in a grammar school; more of the girls will be taking Domestic Science; more boys and girls will be taking the subjects of a commercial course. The G.C.E. and the C.S.E. are not the only examinations taken; examinations of the Royal Society of Arts in Shorthand, and Typing, can be in the programme.

The range of homes from which pupils come may be wider than that of a grammar school. Some comprehensive schools depend almost entirely on new housing estates. More of the parents are paid by the day or the week. Fewer of the parents have themselves known any school life after the age of 14. Fewer of the parents themselves have taken any external examination. Through the success of comprehensive schools in building up their fifth forms, the benefits of further education (in the sense of education continued after fifteen) are being extended down the social scale.

Losses during the course of the fifth year must be expected. The internal examination half-way through the course may precipitate decisions to leave. Many schools are in areas where there is work—of a kind—for nearly everyone who leaves at 15. Seeing friends who have started work with all evenings free and money to spare sets up one temptation, affecting both boy and girl. If the adults at home speak of school as a place to leave at the earliest opportunity, the adults at school who wish a boy to stay in his own interest are fighting a losing battle. Teachers in comprehensive schools have to learn to bear the disappointment of seeing able boys and girls leave even after entries have been completed, leave without the initial qualifications which would help them to climb out of the class and background they inherit.

86

CERTIFICATE OF SECONDARY EDUCATION

In some areas the C.S.E. examination has brought little change because it has replaced regional examinations—the Union of Lancashire and Cheshire Institutes in Lancashire, the London Secondary Schools Examination in the London area, the Union of Educational Institutions in the Bristol area. In other areas it has ended the situation of 'G.C.E. or nothing'. For example, one small school used to enter two forms for 'O' Level. The lower of the two, regarded as of poor quality, managed to average $2\frac{1}{2}$ passes per pupil. Now the pupils in the lower form can take three or more subjects at 'O' Level and the rest in C.S.E.; their quality is no longer measured only by G.C.E. standards. With two standards, success is within reach almost in every subject.

It will take time for the stripling C.S.E. first to earn, then to acquire prestige. A few teachers doubt the readiness of employers to recognize the new examination or to see merit in Grades below the top one. In some areas a boy whose passes in C.S.E. are not recognized may find that he has lost his chance to enter into apprenticeship because he is now too old. The first year's working of C.S.E. has thrown up instances of too many still trying for 'O' Level rather than for C.S.E.—'at their cost'. Previously, 'the prestige of "O" Level attracted pupils who were unable to handle such a course'.

Teachers appreciate their opportunity to shape syllabuses to fit the needs of their pupils. A few have boldly opted for special syllabuses, designed for one school only. Some schools choose to join a small group to draw up a group syllabus. A price has to be paid—loss of teaching time through 'attending interminable meetings'. This is worth paying if the result is (as some claim) 'the introduction of exciting courses'. From another source we learn that 'Some Departments (e.g. English) have found the introduction of C.S.E. a blessing that has added a new stimulus to the teaching'. In English Literature, the set books for 'O' Level (sometimes inviting the accusation that they are drawn up with A forms of grammar schools in mind) may be unsuitable or unattractive whereas the local syllabus in C.S.E. pays much more attention to the interests and limitations of the average pupil of the whole age range. Too wide a gap between G.C.E. and C.S.E. syllabuses in one subject, besides compelling earlier decisions about courses, limits rectification of early decisions by transfer.

The coming of C.S.E. has in comprehensive schools almost ended the distinction in the fifth year between examination forms and forms taking no examination. If it alters the attitude of staff to pupils it must more affect the pupils' opinions of themselves. Several members told us, 'Many more are now viewed as examination material'. It has pushed further away overclouding expressions like 'failure' and 'incapable of a reasonable attempt in the examination'. It has brought success within reach of almost all. Pupils can now regard themselves as capable (as they have been all the time, measured by appropriate standards).

Naturally, the result is harder work, better work. That message runs through these opinions from our members.

(a) 'It is a great incentive to children of average ability to have a chance to take an external examination.'

(b) '...certainly encouraged all the less able boys.'...

(c) 'It gives purpose to pupils in middle and lower streams, giving them an immediate goal.'

(d) 'Pupils are more eager to gain some qualification within their grasp.'

(e) 'Pupils work well for an examination, especially

when they find that it covers their interest and that they can do it.'

(f) 'A greater sense of purpose in staff and boys.'

It is not looking through rose-coloured spectacles to forecast that with the establishment of the C.S.E. there will be fewer determined early-leavers, more boys and girls voluntarily staying at school until they are sixteen. Moreover, there will be more boys and girls, delightedly surprised to find how capable or competent they are, who ask to stay on for a sixth-form course. That will be so, even if there is some truth in the words of one lone voice, the 'wary colleague', with his statement that 'the keen are still keen, the lazy still lazy'. The majority opinion undoubtedly is 'The C.S.E. has had an excellent response from the children in terms of attitude, effort made, and results so far gained'. In the future a report like this will become less exceptional: 'Every fifth-year pupil took at least one C.S.E. subject and gained at least one Grade 4 Pass'. You weigh the force of that 'Every' by thinking of the bottom boy of the whole year.

Combined results

School No. 1 (fifth year under 150)

Subjects passed ...	8	7	6	5	4	3	2	1	0
G.C.E.	6	14	16	16	19	6	12	19	13
C.S.E.	—	26	24	14	5	1	0	18	0

	G.C.E.		C.S.E.						
						Grades			
Subject	Entered	Passed	Entered	1	2	3	4	5	U
English Lang.	114	98	43	0	6	10	23	4	0
English Lit.	86	49							

	G.C.E.		C.S.E.						
						Grades			
Subject	Entered	Passed	Entered	1	2	3	4	5	U
French	51	33	41	3	10	19	6	1	2
German	5	3	—	—	—	—	—	—	—
Scrip. Knowl.	1	0	—	—	—	—	—	—	—
Geography	58	57	60	15	19	11	14	0	1
History	61	42	44	1	8	11	18	5	1
Woodwork	2	2	7	0	2	2	2	1	0
Biology	2	2	—	—	—	—	—	—	—
Chemistry	15	7	13	1	1	3	6	2	0
Physics	75	29	67	5	8	12	23	16	3
Maths	86	69	70	28	15	12	10	3	2
Metalwork	21	16	24	0	3	0	9	8	4
Tech. Draw.	70	39	59	8	16	18	11	3	3
Latin	6	5	—	—	—	—	—	—	—
Art/Craft	10	6	24	3	11	8	2	0	0

Notes. 1 Pass: 9 in 'O' Level, 18 in C.S.E. were entered for 1 subject only. 50 of the 'O' Level candidates came as 'non-selective'.

School No. 2 (fifth year under 150)

Subjects passed ...	7	6	5	4	3	2	1	0
G.C.E. total		27		24	21	19	39	27
G.C.E. 'Non-selective'	2	2	6	10	9	—	—	—
C.S.E. total	1	12	20	19	34	30	22	3
C.S.E. 'Non-selective'	1	11	10	12	25	15	12	3

School No. 3 (fifth year under 125)

Subjects passed ...	7	6	5	4	3	2	1	0
G.C.E.		21		8	15	10	15	7
C.S.E.		32		10	15	16	14	—
C.S.E. passed in all	7	10	8	10	10	15	12	—

School No. 4 small school (120 intake) (fifth year under 60)

Subjects passed ... 5 or more		4	3	2	1	0	All
G.C.E.	21	9	7	13	24	8	29

		Grades					
C.S.E.	Entries	1	2	3	4	5	6
Commerce	3	—	—	1	1	1	—
Principles of Accounts	8	2	1	—	2	—	3
Office Routine	7	—	1	2	1	2	1
Woodwork	7	2	1	2	2	—	—
Metalwork	7	7	—	—	—	—	—
Tech. Drawing	7	5	2	—	—	—	—
English	15	3	4	1	5	1	1
Geography	13	6	3	1	2	—	1
Maths	28	6	7	4	6	1	4
Science	12	1	2	3	1	4	1

School No. 5 (fifth year under 150)

The first comprehensive intake was in 1961. Those who took external examinations began in 1960—secondary modern pupils along with half a form of grammar school pupils.

G.C.E.

Subjects passed	5	4	3	2	1	All	0
Pupils	22	8	12	9	23	16	20

Notes. (1) 'The total number of subject passes showed a 61 per cent increase on those achieved in 1964 by the 1959 intake which was purely secondary modern.' (2) 'The pupils entering the school with 11 + scores approximating to those required for grammar school did not provide the majority of the best G.C.E. results.'

C.S.E.

(1) Maximum number of pupils entering in any one subject— 92 (Maths).

(2) Average number entering per subject (11 subjects)— 11.

(3) Number of pupils gaining Grade 1—33.

(4) Number of pupils with Double Entry—both G.C.E. and C.S.E. in the same subject—who both passed in G.C.E. and gained a Grade 1 in C.S.E.—20.

(5) 'Only one pupil gaining a pass in G.C.E. failed (i.e. below grade 5) in the same subject at C.S.E.'

GENERAL CERTIFICATE OF EDUCATION

We have stressed that C.S.E. results must be rated in the light of the composition of fifth forms in comprehensive schools. Even more background knowledge is needed before one can make a just assessment of 'O' Level results. It must not be forgotten that the successful candidates of 1965 included many who would not have been thought capable of working on a G.C.E. course when they began in the comprehensive school, as well as many who, had they gone to a modern school, might have developed too late to make their way into the only form entered for 'O' Level subjects.

A few years ago the 'O' Level entry in a grammar school was 80 per cent fifth-formers on a five-year course. Since then there has been some increase in the number of candidates from a four-year course and a very much bigger increase in the number of candidates from the sixth form. Apart from second shots at subjects important for University Entrance—English Language, French (for Science students), and Maths and Science (for Art students)—the General Paper is now taken by students from both lower and upper sixth. So in the results of any school entering candidates for 'O' Level, care must be taken not to read too much into the figures for candidates passing in only one subject— a creditable result if they have entered for one only. Nor do the figures for candidates passing in no subject necessarily indicate poor material or entering candidates who should not have been entered. Fuller use of the opportunities presented by C.S.E. has been made in comprehensive schools than in grammar schools. The result is a greater proportion of candidates entered for one or two subjects, and a smaller proportion of candidates entered for their weakest subjects.

Examples of 'O' level results needing explanation:

Subjects passed ... 5 or more	4	3	2	1	0	All
School X	43	14	9	21	15	23	N/A
School Y	36	14	18	36	49	42	47
School Z	32	26	24	78	118	Not given

School X

Entry included:

16 from Form V taking a full range.
10 from Lower VI taking only a few subjects.
19 from Upper VI taking General Paper and single subjects.
20 from Form IV taking a few subjects.

School Y

The entry can be re-presented as follows:

Subjects passed	5	4	3	2	1	0	All
Fifth forms	36	14	15	16	13	10	7
Fourth forms (1, 2, or 3 subjects)	—	—	1	15	9	3	24
Sixth forms	—	—	2	5	27	29	16

School Z

(1) 'Three Boards taken and standards vary.'
(2) 'Departments vary in standards and permissiveness in allowing marginal candidates to entry.'
(3) Some fourth-form entries for one or two subjects only.
(4) 'Many less able fifth-year pupils on mainly C.S.E. work are entered for 1 or 2.'
(5) Some lower sixth pupils, both those intending and those not intending to take 'A' Levels, are entered in new subjects for the first time (e.g. Economics, British Constitution, Economic History).

Further 'O' Level Results—July 1965

Subjects passed ...	5 or more	4	3	2	1	0	All
School H	28	14	14	38	73	106	17
I	5	5	5	10	27	6	1
K	46	25	39	46	49	85	—
L	38	7	17	19	23	5	31
M	8	3	6	20	55	43	24
N	19	15	18	15	22	9	13
O	11	6	10	13	28	22	3
P	26	5	8	18	44	21	32
Q	11	9	13	25	38	36	23
R	23	5	4	6	20	33	30
S	15	3	5	5	2	0	—
T*	22	6	10	10	8	—	—
T†	—	4	7	6	20	—	—
U	19	14	11	19	20	19	11
V	23	8	10	18	38	13	28
W	5	7	6	12	9	18	3

* Form V.   † Form IV.

Notes:

School H	Last year of fast form, now abandoned.
I	Receives from Modern School at 13+.
K	31% of intake have 5, 44% have 3 passes.
M	Fourth forms for Maths only.
N	'If they have done the course, they sit.'
O	First year of 'O' Level.
P	26 lower sixth.
Q	Fourth form for one subject.
R	Small school.
S	Small school. Entered if likely to pass.
U	17 entered for Eng. Lang. only.
V	52 candidates combine G.C.E. and C.S.E.
W	Candidates started at 12 in Form II.

SPECIAL COURSES

In the previous section we referred to good work in organizing relevant and stimulating courses for Newsom children. In a very large comprehensive school the numbers of those who are well advised to limit C.S.E. participation to one or two subjects make it possible to provide coherent courses. There is no question of allowing pupils to fill their time by sitting in classes working for C.S.E.

So in the school whose Special Courses were praised in the previous section, the fourth-form work continues naturally into these fifth-form Special Courses:

Auto engineering	Building
Catering	Commerce
Nursing—Child Care	

Practical Studies continue—Woodwork, Horticulture, Housecraft, Needlework, and in addition there is a practical element in Home Activities (for boys as well as girls). 'This World of Ours' continues as a blend of History and Geography; 'Meeting Point' maintains its new approach to Religious Instruction; 'Money and Machines' still provides unconventional treatment of Mathematics.

Science outside the C.S.E. framework means a course in Rural Science with a nine-weeks section on Bee-Keeping. There is an optional course on Local Studies (directed jointly by History and Geography staffs) which can be a C.S.E. subject, examined on the school's own syllabus. Team-work (in teams of two or three) is the striking point about the course 'Into Europe—an enquiry into the everyday life of people like ourselves in other countries'. Room is made for 'Leisure', the aims of which include 'To help young people to cope with the expected increase in leisure time' and 'To off-set the fact that the majority of leavers of limited ability will do jobs which are unimaginative and monotonous'.

No opportunity of taking pupils out of the classroom is lost. Visits are cardinal and regular features of many courses. Boys and girls taking the Homemaking Course see all helpful exhibitions and study the shops in lesson time. Some courses go beyond short visits. Pupils taking the Commercial Subjects course go into offices for Works Experience, so that while students they can 'experience as real situations the skills and topics they have met in the classroom'. Every boy and girl enjoys 'six separate weeks of actual Work Experience in six different places of employment'. Six weeks represent nearly one-sixth of the teaching time of one school year. Such pioneering work is possible only where the Local Authority gives unstinted support, where local firms give whole-hearted co-operation, and where there are determined organizers refusing to be beaten by technical complications such as questions of insurance. In a school like this the last year at school is truly a bridge between the small world of school and the large world around.

9

THE SIXTH FORM

Those in comprehensive schools who teach or organize sixth forms must often feel that they are performing in public, that eyes are fixed on them. If the eyes are working in conjunction with a mind already fixed in the idea that these schools are a mistaken experiment, evidence will be seen that supports the preconviction. If with the eyes goes an open mind, enough will be seen that is material which is to be weighed in the balance. Where the mind has decided in advance that there can be nothing wrong with these schools, there is plenty for the subservient eyes to feast on. Success, like Beauty, lies in the eye of the beholder.

Unfortunately, the laziest way of measuring the success of sixth forms is an unsatisfactory one, an unfair one. Counting up passes at 'A' Level is a poor way of measuring the success of any school with a sixth form. Figures are easy to produce and easy to exploit. If you could line up the students who leave the sixth form at the end of the second year in that form, you could see the number and kind of persons going out into the world. If the quality is good, the nation will benefit wherever they go for the next stage of their education. If you could prove that too few go to universities, you have done nothing to diminish the success of the school in bringing such students up to university standard. You would need to be very manipulative, to find any 'proof' in the figures we give here. Now teaching in comprehensive schools are graduates who as boys were pupils in such schools. The numbers recorded as proceeding to a training college may be as sound a basis for pride as those in any other column. The figures for one school include two who began in Remedial Classes.

The success of a sixth form is to be seen in the daily working of a comprehensive school. Throughout the school many pupils have a target—to reach the sixth form—which provides an incentive and encouragement to continued effort. Throughout the school, interest in and participation in the very many clubs and societies is increased because of the liveliness of the students in the sixth form. Where House spirit is strong and beneficial, a good sixth form is usually a major factor.

The benefits of being in a sixth form are, in a comprehensive school, open to a wider range of pupils than in any grammar school. A consensus of opinions expressed at one of our conferences was: 'All the traditional methods used to develop personal qualities, encourage judgement, etc., could be applied successfully to many more pupils than previously. It was felt that the comprehensive school would probably do this better than the average grammar school because of its size and the consequent wider variety of opportunities.' Even in 1966 there will be in grammar schools pupils whose decision not to continue their studies in a sixth form is based on 'I don't think I'm good enough'. In comparison, a comprehensive school's sixth form is an open society. If you're old enough, you're good enough. Being in a sixth form is more of a natural completion of continuous education. There seem to be fewer losses of able pupils in a comprehensive school. A grammar school may lose some because pupils, no doubt wrongly, think it is a matter of passing so many 'A' levels and because not every grammar school can cater fully for those for whom the right programme is not the usual groups of three subjects on the Science side or three subjects on the Arts side. A modern school may lose some able pupils because they dislike the idea of changing schools before they can continue their studies at a higher level.

The variety of entrants to a sixth form must be appreciated before justice can be done to the results in external examinations. A lower sixth form in a comprehensive school is first and foremost the sixth year, not necessarily a year beyond 'O' Level. It brings together those of the same age with all their intellectual disparity so that together they can be 'treated as young adults and given more responsibility in the management of their affairs'.

A lower sixth form may include:

(1) Students intending to stay for one year only
    (a) to take 'O' Level subjects again
    (b) to take a Commercial Course
    (c) to take 'O' Levels after passing in C.S.E.
    (d) to take 'O' Levels for the first time at the end of six years instead of the more usual five years
    (e) to take one 'A' Level (usually Art) at the end of the first year in the sixth.

(2) Students from other schools:
    (a) 'from secondary modern schools, specifically for the chance to take the variety of subjects we offer' (transfer before 'O' Level)
    (b) from modern schools which have no or little post-'O' Level work
    (c) from neighbouring grammar schools which operate an 'O' Level hurdle before allowing fifth forms to move up to the sixth.

It is to be expected, then, that every form in the fifth year of a comprehensive school will be represented in the sixth form. There is a combination of those who in other areas would have gone to a grammar school and those who in other areas would have gone to a modern school. How can one compare the results in an external examination with

those obtained in a grammar school? Take these results
of the second-year sixth in one comprehensive school

Grades	A	B	C	D	E	Total
Numbers	11	24	6	10	15	66

It so happened that the 1958 entry in that school was
subject to an 11-plus division. So the figures can be re-
arranged

Grades	A	B	C	D	E	Total
21 passed 11+	6	14	4	4	10	38
16 failed 11+	5	10	2	6	5	28

Of these 37 candidates, 7 proceeded to a university.

Sometimes the Intelligence Quotient has been determined
and is known. In assessing the sixth form of one school, one
has to hear what the teachers say: 'As virtually every pupil
with an I.Q. of 130 and upwards is creamed off, we lose
20–30 members of the potential sixth form.' Another
school knows that half of its sixth-form students had an
I.Q. of under 110. Yet 16 passed in three subjects at 'A'
Level, seven entered for two subjects and passed in two.
Moreover, 'No candidate failed to obtain at least "O" Level
in each subject taken. No candidate failed to get a place in
the type of further education for which he was working.'

By grammar school standards, a sixth form in a comprehensive school includes many who could be called 'non-
academic'. The disparaging label covers many able children—'academic' is only one direction. It is something of
an achievement that in a comprehensive school a division
into 'academic' and 'non-academic' can be avoided.
This suggests the basis of the different approach. There may
be backward grammar schools where the organizers start
with courses and fit children into them. Certainly the com-

prehensive school starts with children and offers and supplies courses to fit their needs. 'It is our job to find the right kind of course for every one', said one of our members at a conference. Mostly they succeed. Early enquiry before the end of the fifth year determines the courses asked for. If it is possible they are provided in the *ad hoc* time-table. Indeed it comes very near to working an individual time-table for each student. It is easiest with the really able boy who can take three subjects at 'A' Level in his stride. It is hardest with those at the bottom of the scale—'All too often pupils of poor ability wish to stay on and have to scrape around for subjects to study, giving rise to unusual combinations'.

At our conferences we heard that a programme combining subjects at 'A' Level and subjects at 'O' Level was very much a normal and expected thing in a comprehensive school. Where possible, the time-table is drawn up to allow for such combining of levels. Our members were firm in their belief that it was easier so to combine in a comprehensive school than in a grammar school. Many schools take great care with explanatory sheets for students and for parents, setting out the choices available in the sixth. One such circular states bluntly 'Many candidates are not capable of taking three subjects at "A" Level'. A good circular expounds the importance of finding the course that is appropriate. There is no suggestion that the student with a two-subject range at 'A' Level is somewhat inferior, that a student with a one-subject range is quite inferior to a student who can and should tackle three subjects at 'A' Level. Everything is done to ensure that no one starts in the sixth form with a feeling that he is a misfit. There is no suggestion that there is a standard course, that students of widely different abilities should bear the same burden or proceed at the same pace.

The larger the sixth form, the easier it is to follow the usual practice of not making a division into Science sixth and Arts sixth. There is greater expectation of and provision for a combination of subjects that cuts across the division. A pattern of English and History and Mathematics is not unusual. Many schools are able to arrange that Geography may be combined with subjects from either Arts group or Science group; fewer schools are able to do this for Economics.

Our members feel that a comprehensive school is better equipped to enable a student to combine the 'A' Level subjects he needs for his further education with the 'O' Level subjects he needs to satisfy Entrance Requirements or to continue the subjects he is interested in. So in the lower sixth, a student is more likely to be able to study for 'O' Levels in the supplementary examination (November or January) alongside his 'A' Level subjects. If such a student prefers to or needs to take 'O' Levels at the end of the first year, that too can be arranged. Even in the second year of the sixth, the time-table can be arranged to allow for 'O' Levels —from the start. Much use is made of the opportunity to start new subjects in the sixth, taken at 'O' Level at the end of two years. Courses are less determined by what boys did up to the fifth form. A switch is possible in the sixth form. For example, boys who had chosen other subjects in options in preference to practical studies can now start on Metalwork and Technical Drawing. Those who were rather late in showing ability in languages can now start a second language—often German. Scientists can begin Russian in schools where there is a teacher. As a demand for something new or something different arises in order to provide an appropriate individual time-table, there is in comprehensive schools a determination to satisfy the demand. Usually the means of satisfying the demand are there, too.

Interesting experiments are taking place in providing for the first-year sixth a coherent one-year course in secretarial training, not necessarily geared to external examinations. In one school 32 periods per week are given to Shorthand, Typing, Book-keeping and Commerce, with eight periods of minority studies and optional activities. Another school puts more emphasis on broadening the outlook in preparation for entering the business world. There, in addition to Typing, Shorthand, and Commerce, students attend lessons in Art, English, Greek Civilization, and Human Biology. Where the syllabus of an 'O' Level examination is helpful and appropriate, the students take 'O' Level.

Groupings

In the summer term it is usual for a fifth-former who intends to stay on to be given a schedule setting out the subjects from which he is to choose to find the right course. Examples of such lists are given below. They are samples of September 1965, not permanent fixtures. We are reminded by those in the schools concerned, 'These lists are altered each year to accommodate the wishes of intending sixth-formers'.

One subject to be taken from three of four groups

(The number of subjects taken at 'A' Level may be 1, 2, or 3)

English	French	German	Music
Botany	Applied Maths	Zoology	Geography
Physics	Chemistry	History	Pure Maths
Economics	Technical	Housecraft	Pure and
Metalwork	Drawing	Geology	Applied Maths
		Accounts	

One subject from each of three or four groups:

SCIENCE	Maths	Physics	Chemistry	Biology
	Pure Maths		Applied Maths	Technical Drawing

ARTS	French	History	English	Geography
	Scripture		Geology	Latin
	Art			Music

One subject from each pool, with arrangements for two subjects at 'A' Level, two at 'O' Level

Pool 1	Pool 2	Pool 3
English	History	Technical Drawing
Maths	German	Geography
Biology	Music	French
	Physics	Art
	Biology	Chemistry
		Scripture Knowledge

One group to be chosen

History	Physics	Physics	Geography
Economics	Maths 1	Chemistry	Woodwork
Geography	Maths 2	Biology	Technical
or English			Drawing

A testing point is presented when the student can choose only one subject from a group or pool. The more items there are in the particularly mixed group, the greater the chance that some student desires to choose *two* of the group. If such a group includes

History	Economics	Chemistry
Art	Maths II	Domestic Science

there will be girls whose first choice is Domestic Science and whose second choice is History or Economics. Some boys will feel frustrated if they are barred from doing *both* History and Economics. Another example of a 'last group' with inherent complications is

Chemistry	Technical	Geography
Latin	Drawing	Music
	Art	

The boy who particularly wants Technical Drawing may well wish to do either Chemistry or Geography. Students whose first choice is Art or Music do not fall into any standard pattern. Their second choice may be from any of the subjects normally taken by the Science sixth or the Arts sixth or the Commerce sixth of a grammar school. No doubt special arrangements will be made for special cases.

It is a useful exercise to study the full programme of courses offered in one school (under 750 pupils).

Lower Sixth

'A' Level Courses   'O' Level

Arts	Science	G.C.E.
English 6	Physics 7	Economics and Public
History 6	Chemistry 7	Affairs 4
French 6	Botany ⎫ 3+7	Russian 2
Latin 6	Zoology ⎭ with U6	
Geography 6	Single Maths 4+4	
	(with Double Maths)	
	Double Maths	
	4 Applied	
	6 Pure	

Art—2 periods—with U6
Technical—6 Woodwork 6 Technical Drawing

Upper Sixth

Arts	Science	General
English 7	Physics 7	Russian 2
History 7	Chemistry 7	
French 7	Botany ⎫ 2+7	
Geography 7	Zoology ⎭ with L6	
German 6	Single Maths 4+4	
Art—2 periods	(with Double Maths)	
(with L6)	Double Maths	
	4 Applied	
	7 Pure	

105

General Courses	Upper and Lower Sixth
Practical Aesthetics	2 periods in forum with 4 staff
History and Philosophy} *or* Art and Science }	3 periods in forum with 4 staff
Religion and Philosophy	1 period
Games	2 periods

Problems

Sixth-form work brings its own problems, some of them deep-rooted and long-lasting.

(1) The locality. A school may be in an area without a tradition of continuing education after the age when the law allows children to leave. Easy chances of immediate employment may influence parents and deter an able boy from continuing at school when his friends have so much money to spend. The locality may make it difficult to have a balanced sixth form; more than one school reports being strong on the Science side, weak on the Arts side. Again, some areas have built-in attractions which make recruitment of staff easy for them, so that the area that is thought unattractive may have to take what offers when there is a vacancy. It is harder in such 'unattractive' areas to retain staff; lack of continuity may be more harmful in sixth-form work than lower down the school.

(2) Accommodation. The more successful the school is in providing tailor-made individual time-tables, the greater the call on rooms, for teaching small groups, for tutorials, for private study. Schools which succeed in developing a strong sixth form quickly have greater numbers than the planners provided for. It takes time and drive to get temporary additions, much more so permanent extensions. It is seldom feasible to restrict the entry to make more room at the top.

(3) Transfers. Some of the pupils who come from modern

schools are slow to settle in so that a three-year course to 'A' Level may be necessary.

(4) Training. Even if form-masters take deep interest in their charges and even with a tutor system, it cannot be left to chance for weaker students to learn to plan their time and use their time. It is not easy to find a happy medium between too strict supervision of private studies and too little.

(5) The growing school. It takes time to build up a staff that can provide a full variety of courses. In the early stages, there is often only one teacher for each subject with experience in teaching both 'O' Level and 'A' Level. This may give him a very heavy time-table. There may not be enough to offer in a subsidiary modern language such as Spanish, Italian, to attract a true specialist. 'The first year of 'A' Level is a testing time for the *Staff*', one correspondent tells us—because for some of them it is their first year of 'A' Level preparation. Where the staffing-ratio is weighted in favour of sixth-form pupils, there is a time-lag before the increase in the number of teachers so allowed becomes operative.

(6) Restrictions. Almost all schools report that there is no minimum number of students before a sixth-form class can start. An exceptional case is the school which demands three students before a class is arranged.

(7) Pressures. Demands for sixth-form classes are often met only by sacrifices.

> (a) In a few schools a minority group will have a lesson time-tabled for Period 0 or Period 9. Translated, this means that the teacher is putting a lesson in before the first lesson starts or after the last lesson ends—in his free time. The result is an overlong day for both teacher and taught.
>
> (b) Combining first- and second-year sixth classes—

'The first and second year are taught together in all but Maths.' Such doubling, unfair to all concerned, is unfortunate.

(c) Staff giving up free periods. 'Where staff is not, strictly speaking, available, Heads of Departments have often sacrificed periods voluntarily rather than say "no" to a keen pupil.'

(d) Effect on the Middle School. Because of sixth-form needs, fewer of the third and fourth year in a small school may have graduates to teach them. One school ended the doubling of lower- and upper-sixth Science only by making Science optional in the fourth and fifth years. Another school reports difficulty in providing a variety of courses in the Middle School as well as meeting sixth-form demands because their resources were insufficient for both.

(e) Workshop subjects. Though the numbers taking Advanced Woodwork, Metalwork, Technical Drawing may be small, each requires the use of a specially-equipped room for six to eight periods per week. This makes it more difficult to cater for lower forms which may have to be sub-divided because a workshop cannot accommodate more than half a class in a boys' school.

(f) High-fliers. Very, very few comprehensive schools operate a fast form to bring its ablest (academically) pupils to 'O' Level in four years. This has less to do with size than with belief. There was general agreement at one of our conferences that it is better to give an all-round education over a broad range for five years to 'O' Level than to start early specialization by means of a four-year course with restrictions in subjects. It was believed that the high-fliers are better for not getting to the sixth form too soon. One school whose representatives spoke strongly on this point gets its share of successes in places at Oxbridge. Some schools put high-fliers in for 'A' Level as soon as they are ready —at the end of the lower sixth or in the January of the year in the upper sixth.

(g) Modern languages. A born linguist should choose his school carefully. Some schools find difficulty in providing staff and arranging lessons for modern languages to be added to French. Occasionally such a boy may have to choose between a cram course with the hope of 'A' Level in two years and a course begun in the sixth with 'O' Level in two years.

(h) Classics. Some schools are ready to arrange classes of one or two in Greek and in Ancient History and Literature. In another school no Classics at all. In another school, 'Latin is taken to "O" Level, Greek not at all'. In another, 'Only one pupil for Latin, taken in free periods by the Head of the Languages Department'. In several schools, 'Arrangements are made for Latin "O" Level for University Entrance'—i.e. Latin cannot be started until the sixth-form stage. Where there is no Latin teaching in the fifth form and below, it is most unlikely that there will be on the staff a graduate in Classics. The number of students affected may be small but provision for the boy with latent ability in the ancient languages is maybe the severest test of the claim that 'If the demand is there, the comprehensive school can supply the need'.

(i) Difficulty of catering for Botany, Zoology, Economics, Scripture Knowledge at 'A' Level.

General studies

This Association's booklet *General Education in Grammar Schools* referred to the way that comprehensive schools were fully aware of the needs and were trying to maintain a balanced programme in the sixth form. All the evidence supports the claim that now General Education is respected and provided as well in these schools as anywhere else. One school working on a time-table of 70 periods per fortnight provides 48 periods per fortnight for 'A' Level studies and 22 periods per fortnight for 'Minority time' or general studies. Different ways of ensuring that the students are more than mere adepts in 'A' level subjects include:

(*a*) A first-year sixth form has two *afternoons* each week in House Groups for Liberal Studies.

(*b*) All the lower sixth have periods for English, Social Studies, Religious Instruction, Physical Education and, in addition, a Group period.

(*c*) All the lower sixth have four periods for English, and Music for two periods is optional.

(*d*) Courses in History of Human Thought (one period), History of Scientific Thought (two periods).

(*e*) All the sixth have Art and Music Appreciation besides periods with the Headmaster, the Deputy Headmaster, the sixth-form Master.

(*f*) All Arts students have two periods of General Science; all Science students have two periods 'Humanities'.

(*g*) Periods specified for 'Use of English' in time-table for the upper sixth.

By now, careful provision for studies outside 'A' Level subjects is so well established that many schools regard it as too common a feature to be worth special notice.

Combined efforts

Our correspondents were asked to give their views on co-operation with neighbouring schools in providing sixth-form work. Most schools saw no need for it and little value in it. Some schools thought the disadvantages outweighed any advantages. One period spent in a neighbouring school might (because of travelling) take up three periods. A serious practical point was raised—to be able to co-operate, two schools must have the same number of periods per week and, preferably, similar times for starting lessons. A school with a 70-period fortnight could hardly keep in line with another school with a 40-period week. The smallish school in a rural area may have the greatest difficulty in providing the variety it thinks desirable, yet the thinly-populated area it serves most likely means that the nearest school with sixth-form work is too far away to allow co-operation.

In a city with several comprehensive schools such co-operation is possible. It is being tried in one city. Two schools there combine forces for

Latin	German	Music	Scripture Knowledge
	Metalwork		Technical Drawing

It can be arranged so that the staff of one school always has sixth-form teaching in each subject, but only one year at a time. The school that has the lower sixth in a subject for 1964–65 retains the students as upper sixth in 1965–66. In the latter year, the partner school starts a lower sixth, which it retains as upper sixth in 1966–67. Presumably a third-year student would count as upper sixth.

Sixth-form units

The word 'College' exerts a fascination on some. In cities it happens that at the end of the 'O' Level year a student can stay on in the sixth or do similar work in a technical college for the same external test. A youth can deceive himself into thinking that transfer to a 'College' is promotion. He may be tempted by the seeming increase in freedom—no restriction on smoking, no uniform, less restraint all round. Our members in comprehensive schools firmly believe that it is in a youth's own interest to stay at school, where he can receive training in study, where he can have pastoral care. They are equally firm that for him to stay is of benefit to the school. They believe that the school by providing opportunities for leadership can help the individual student to realize his own potential.

It will be understood, then, why our members are rather suspicious of the advocacy of sixth-form colleges—for which, they admit, there may be a place in a thinly-populated area which has to have several small comprehensive schools because of distances involved, where no one school can build up a strong sixth form. To our members, taking away the oldest students would be robbing the teachers of a reward earned by hard labour. It would lower the comprehensive school in the eyes of parents and encourage short-sighted pupils to think of the school as a corridor leading to something bigger and brighter, rather than as a first-class unit for providing continuous education from eleven to eighteen. A lopping of the top layer would make teaching in a comprehensive school much less attractive. Nor is the chief factor the reduction in pay which must follow when the unit-total of the school falls. It would be a sad day for the teaching profession if the feeling arose that sixth-form colleges were where the best teachers went and comprehensive schools were where others went. A teacher, it is felt, benefits when the teaching is in depth, roots in the first form and flower or fruit in the sixth.

Apart from any such competition, the near future may see

more attempts to make a sixth form offering almost all that a college can provide—except the name. There is a valuable prototype in one school's sixth-form House:

Since Sept. 1963 a VI Form House has functioned as an independent-organization House in its own self-contained block. It is designed to cater for 3 main categories of students:

(i) Orthodox 3 × 'A' Level Students.

(ii) Those aiming at 6–8 passes at 'O' Level with 1 or 2 passes at 'A' Level.

(iii) Those taking six years for 'O' Level in 4–6 subjects. The Sixth includes 43 in 6G, none of whom have more than 4 passes at 'O' Level. 14 of them have started 'A' Level courses and if they are sufficiently successful in the supplementary examination they will join the lower sixth proper.

Organization

Normal school rules largely cease to apply.

Considerable degree of freedom in dress. (Most choose to remain in school uniform.)

Coffee-bar at break.

Self-service cafeteria system at lunch-time.

Self-signing-in system in lieu of calling registers.

Discipline designed on simple basis of self-responsibility.

Elected VI Form Council.

No locks or keys—mutual trust and confidence.

Table 5. *Sixth-form figures*

School	Sept. 1965 Total number of pupils (to nearest 10)	Sept. 1965 Fifth Form	Sept. 1965 Sixth Form			Pupils in Sixth taking no 'A' Level	July 1965 Passes at 'A' Level [i.e. previous school year's pupils]			1964 and 1965 Average number of students proceeding to			School
			Year 1	Year 2	Year 3		3	2	1	Univ.	C.A.T.	College of Education	
1	890	83	29	22	0	0	11	8	8	8	2	7	1
2	510	90	26	21	0	0	7	9	3	8	1	5	2
3	600	55	33	29	6	8	5	4	6	4	1	3	3
4	600	69	23	25	2	3	5	2	9	3	1	3	4
5	710	†	24	27	4	0	12	6	8	7	2	—	5
6	980	107	52	33	3	21	11	7	8	7	2	6	6
7	1,840	228	138	71	13	75	21	20	8	23	3	4	7
8	1,580	226	69	45	7	8	26	23	7	18	10	20	8
9	2,030	256	156	81	42	25	28	22	11	22	4	6	9
10	1,660	124	52	23	1	27	4	3	3	2	1	2	10

11	1,340	146	42	34	3	14	15	5	5	6	5	1	11
12	1,360	180	73	39	1	25	14	13	8	11	0	6	12
13	1,330	194	114	50	18	58	16	20	21	16	†	†	13
14	970	127	64	24	3	22	15	5	6	9	3	6	14
15	1,200	166	42	28	1	25	3	10	5	1	1	4	15
16	810	74	41	21	5	12	6	12	5	6	4	7	16
17	950	124	40	14	0	15	5	9	2	1	5	7	17
18	960	137	64	40	3	15	6	6	8	7	1	5	18
19	960	106	53	36	5	†	14	12	15	12	1	5	19
20	1,920	226	70	25	7	20	17	18	20	16	2	11	20
21	1,380	137	43	23	4	12	11	10	12	7	0	6	21
22	1,940	161	42	20	2	0	16	8	6	6	3	7	22
23	910	107	40	18	2	16	6	1	6	2	0	5	23
24	1,450	145	61	48	15	12	19	20	11	12	2	15	24
25	800	134	11	16	5	†	7	7	14	6	3	2	25
26	1,220	141*	26	26	0	0	11	9	10	6	4	2	26
27	1,650	187	19	10	3	0	1	3	7	1	1	4	27
28	1,300	146	32	20	1	0	9	1	2	5	0	1	28
29	1,110	160	38	18	1	15	5	4	11	3	4	5	29
30	1,420	130	22	11	1	0	7	5	7	3	4	3	30

* 61 in 6T = 'O' Level form. † Not available in September.

10

OUT OF CLASS

CLUBS AND SOCIETIES

The organization of activities outside the classroom is something in which teachers in comprehensive schools can take
pride. These teachers are to be congratulated on the efforts
they have made and on the results flowing from such efforts.
More than one assistant master moving from a grammar
school to a comprehensive school has been 'impressed by the
range of societies and the number interested'. If comparison
with similar activities in grammar schools is insisted upon,
it must be said that comprehensive schools do as much,
if not more, and in face of bigger obstacles.

The way the school is built can, and too often does, make
it difficult to have active, well-supported school societies.
Spread is a bigger enemy than size. In a grammar school
the pupils who attend a meeting at 4.15 p.m. of the Stamp
Club in Room 18 probably come from their form rooms
within one building. In a comprehensive school, those who
assemble for a meeting of the Stamp Club at 4.15 p.m. may
come from five other blocks of rooms. Little things can be
big troubles to younger pupils—the posting up of notices in
many different places, the disposing of raincoat and school-
bag. Where there are separate buildings for Middle and
Upper School, for example, it is harder to bring together interested pupils from both. It benefits both older and younger
pupils if the sixth formers meet fourth formers in school
societies. Several members regretted the handicap of poor
design, of lack of design in the layout. One member felt
frustration through 'the sprawling nature of the campus'.

Distance is a troublesome enemy. A pupil in a comprehensive school is likely to have further to travel to reach home than if he had gone to a modern school. In a rural area, a pupil may have more miles to travel to reach home than is often found in an urban grammar school. As one member told us, 'A rural school is *deplorably* at the mercy of the school bus which deposits pupils at 9 a.m. and removes them at 4 p.m.' A boy keen on a society not only has more trouble using service buses (involving connections and waiting for connections) but, too, has to pay out of his own pocket the cost of getting home. Those pupils who are provided with a bus pass are luckier. Not all schools have succeeded in making arrangements for school buses to leave school later, say 5.30 p.m. One school which was allowed this arrangement reports that after the change a quarter of the pupils attended clubs and societies on the two nights when the buses left later. Luckier still is the very large, very forceful comprehensive school in which 'School bus' means not a bus belonging to a transport company or committee of a borough but a vehicle owned by the school and used for transporting scholars as required. More than one school has shown its enterprise in this way.

Part-time employment keeps some pupils from joining societies. This affects girls more on Saturday mornings but boys more in the period after 4 p.m. In schools which serve a new town there are more boys on paper-rounds and doing other part-time jobs. It is accepted that sometimes the additional money is needed in the home and that sometimes money so earned turns the scale in the school's favour when the boy has to choose between leaving at 15, to become a wage-earner, and staying on at school, with a supplement from evening work.

Competition handicaps schools in one or two areas.

Youth clubs may flourish at the expense of school clubs. It is admitted that in a city with more than one comprehensive school, the mixing of youths from different schools can bring its own rewards but the habit of going elsewhere for social pleasures operates against the success of school functions. In one city the attempt to time-table most activities for the period between 4 p.m. and 6 p.m. has been abandoned. Instead, the school is opened on four week-nights from 7 p.m. to 9 p.m., and with good results.

No comparison of social activities in two different comprehensive schools would be valid unless the areas feeding the schools were the same. There can be a strong local feeling for or against school activities and school functions. One member writes, 'It is symptomatic of this area that all school functions are poorly attended by pupils and parents'. One area will have a much larger percentage of pupils whose parents in their own schooldays at grammar school absorbed the tradition of school societies. There are parents who will not themselves come to anything at the school and who do not see why their children should want to go back or stay after the end of the last lesson of the day.

A comprehensive school contains children who in the absence of such a school would have gone to a modern school. Every modern school has its pupils who cannot be persuaded to join any club or society. One member expresses the position thus: 'However many opportunities you provide, the lower the I.Q. the less these opportunities are taken up'. Similarly, the more a comprehensive school gets its share of pupils of high intelligence, the fewer it loses to neighbouring schools of the top 5 per cent of the intelligence range, then the more leaders it has with strong home support for active participation. In the words of one member, 'The successful involvement of all pupils in the social life of the school depends upon the strenuous efforts of the staff. Where there is keen interest amongst members of the sixth form, school societies flourish, and this in itself tends to encourage participation by all sections of the school community.'

Where sixth-formers are good channels of communication of the zeal of the staff, school societies most certainly flourish. One school of just over 800 pupils reports 30 different clubs and societies. In bigger schools, we are often told, they are 'too numerous to detail'. Multiplicity may bring conflict of interest in the individual pupil. The number of societies can expand, but not the number of weekdays so that as the variety increases, so does the chance of two interests of the one pupil having their activity on the same night. Drama and music arouse interest everywhere and are the most successful in bringing together pupils of very different ability. It is true to say that practical activities (work in wood, in metal, work on engines and machines) are to be found in greater number and variety than in grammar schools. Equally naturally, there is a little less activity in comprehensive schools in societies based on one non-practical subject.

One school provides parents with a booklet of information about its work and so makes them aware of the opportunities provided. The 27 clubs offered are:

Angling	Dressmaking	Pottery
Art	Flower	Puppetry
Car	Arrangement	Recorders
Caving	Golf	Scottish Dancing
Cercle	Orchestra	Trampoline
Français	Pathfinder	Verse Speaking
Chemistry	Gymnastics	Sculpture
Chess	Judo	Soft Toy
Choirs	Madrigal	Stamps
Drama	Photography	

That list tells much of the interests of the staff. Therein lies both strength and uncertainty. The man who replaces, say, a Maths master may take over the Maths, but may not be interested in Cycling Proficiency; then one club lapses. Indifference is 'combated when a keen teacher organizes the activity but these keen men tend to get promoted quickly'.

DUTIES AND RESPONSIBILITIES

'The Prefect System' is often regarded as a main and permanent feature of grammar school life, so much so that many expect to find it reproduced in comprehensive schools. This is unreasonable, particularly with this newer kind of secondary school which refuses to be limited by the past. It ignores, too, changes which are taking place in grammar schools. In many grammar schools you will look in vain for senior pupils wearing prefects' badges or prefects' ties or other insignia of office. Nor are the schools in a state of chaos: order still prevails.

At least one of the newest comprehensive schools has made a start without prefects. Its representative tells us: 'We have no prefects. Committees, whose membership is open to any pupils, do much of what is asked of prefects in other schools, but there is no hierarchy, a steward's authority being limited to his or her particular job at any given time for a given area.'

This opposition to a hierarchy is found in other schools, where there is determination to prevent a gap between the superior few and the others. The gratifying result is that all senior pupils have opportunities to exercise leadership. In another school it works this way: 'The sixth form elect a committee which organizes the duties of prefects. The school is not run on the lines of a hierarchy. Neither for its discipline nor for opportunities for service does it depend on a prefect system.'

It has been felt in quite a few schools that academic success tends to exert too much influence when prefects are chosen. Two things are desired: first, that the pupils chosen should be the ones with most fitness for the job; secondly, that all grades of intellectual ability should have opportunities according to their qualities of character and power of personality. As one member puts it, under the old way of running a prefect system, 'As time goes on, the pupil who would have risen to the top in a secondary modern school will find it much less easy to become a school prefect'. Another school which has retained prefects but avoided restriction tells us: 'The less academic pupils fill these posts quite often—in fact we have found that they are far more willing than the academic pupils'.

So, few schools have full school prefects drawn from the sixth alone. One of the minority representatives writes: 'School prefects are always sixth formers, thus the less academic pupils who never reach the sixth form never fill positions of responsibility as prefects'. Another school draws its prefects from the *lower* sixth: 'The upper sixth are relieved from these "menial" tasks and provide services to the community'. Still another school goes down as far as the fourth year when choosing prefects, taking the best people, not the brightest only.

Other ways of organizing duties include:

(I) In schools which have separate Upper School, Middle School, Lower (or Junior) school. Each sub-division has selected pupils operating inside that sub-division only. They may be called 'prefects' in the Upper and Middle divisions, but 'monitors' in the Lower (or Junior) division.

(II) In schools with separate buildings for separate Houses.

(a) 'Since the school is broken down into Houses, there

are opportunities for children to take on responsibility either on a House or a School basis. There are House and School captains, House and School prefects, and, indeed, most children hold some post of responsibility before leaving school. School prefects are usually appointed from the sixth form, while House prefects are chosen from the fourth year and fifth year.'

(b) 'Prefects and Monitors are chosen by Houses. Each House is responsible over a period of 9–10 weeks for the control of a given area of the school and/or playground. House Prefects can qualify for promotion to Senior House Prefect or School Prefect. Each House has its own Head Boy, Head Girl. House Monitors are chosen from the fourth form.'

(c) 'House Prefects chosen from third and fourth years. As it is assumed that the abler ones will reach the sixth form and have opportunities there, the pupils from middle streams are mostly chosen. Prefects come from the fifth year. They hold weekly meetings on policy and progress. The sixth-form pupils automatically have prefect status, as "responsible senior citizens".'

(d) 'House Monitors are appointed by House Staff from the fifth form. Their duties are confined to the House area, including assisting the House tutors. Each monitor is attached to a tutor-group and supervises the Tutor-room on wet days. House Prefects are appointed from House Monitors of at least six weeks' standing. A House Prefect is required to make a declaration to fulfil his duties, this in front of the whole House. House Prefects of at least one term's standing may be appointed School Prefects (making a similar declaration before the whole school).'

Everywhere there is evidence of a desire to bring as many pupils as possible into positions where they can contribute to the smooth working of the school community.

One member goes so far as to say that if the Headmaster believes in comprehensive education he must lean over backwards to concoct duties and positions for those below the sixth. So in some schools there are monitors for almost everything. One school has 'Warden's assistants'. Several schools have bus prefects—chosen from below the sixth form.

Experience of representing others is given in several comprehensive schools. Just as there is a Staff Council in some schools, so there can be a School Council, of pupils with staff assistance, on which each form is represented. With a separate Junior School there is sometimes a Junior School Council. One school calls its representative body a Service Committee. All of this is good training for responsibility, for decision-making, for a quiet form of leadership.

Only mention can be made here of what is dealt with more fully elsewhere—the frequency of visits, all presenting opportunities for service and responsibility. Even a one-day outing as an extension of classroom studies of Geography needs leaders and helpers. Much more opportunity is presented when a school takes over the premises of a disused railway station as a centre, or has its own centre in North Wales. One school organizes House Camps which can at any one time take half of the members in a House. In schools like these it is remarkable if a boy can stay even five years and not find an opening for his interests and ability, not develop his character and leadership.

11

THE TEACHER

In 1959 members attending our conferences impressed on us that a comprehensive school is as good as its Headmaster. We would amend that in the light of our 1965 conferences to 'A comprehensive school is as good as its leadership'. The leadership now more clearly includes those who rank as Assistant Masters but who might well be called Assistant Headmasters. This does not stop at the Deputy Headmaster and Senior Mistress (or Senior Master, where the Deputy is a woman). It brings in Heads of Houses in House-based schools, Head of Middle School, Head of Lower School in horizontally-divided schools. Looking ahead, we see many fewer opportunities for Assistants to become Headmasters and more opportunities for men of Headmaster calibre to undertake responsibility and to direct operations in senior posts in comprehensive schools, as Headmaster-equivalents.

There are now many more men who have added experience in comprehensive schools to basic training in grammar schools, men who have played a vital and possibly under-recognized part in establishing comprehensive education as a workable way of organizing education. Where grammar school standards and traditions have been absorbed into the comprehensive way, much of the credit must go to them. There are, too, younger men of high quality, with little or no grammar school experience, who have deliberately chosen to teach in comprehensive schools because of their beliefs. Both classes of teachers have much to give and comprehensive schools have gained very much from them. Figures on p. 14 show that in September 1966 there are more comprehensive schools than in September 1965. In September 1967 there may be very many more than in September 1965. The problem may be akin to that of rapidly expanding an army: how and where can you find all the trained and experienced senior officers that are needed without leaving the original smaller body drained of its leaders?

Many teachers in grammar schools who read this will be in the position that soon the school in which they serve will become part of a comprehensive school. For them, the gates of such a school might carry the words 'Abandon prejudice all ye who enter here'. A comprehensive school cannot do justice to all its children if it tries to imitate any one kind of school with a different intake. In the following pages we look at some of the essential differences which the teacher starting service in a comprehensive school must accept if he is to adjust himself. After all, there will be fewer havens for the unhappy teacher to escape to.

In a grammar school already established in 1900, governors will be influential and important, felt to be part of the government of the school. In the newer maintained schools governors are more in the background but still part of the scene. In the comprehensive school, governors are still more shadowy, still people of considerable power who can take as much or as little active interest in the staff as their nature directs. One school speaks of its governors as remote—'which may be a good thing'. Another school sees them as 'functional only'. In another area, 'We share our governors with other schools and therefore they have no special interest in us'. A different school correspondent writes, 'I see no difficulty in the staff submitting memoranda or complaints direct to the governors if they wish to do so, either collectively or individually'. The best relationships with governors come when efforts are made to see the governors not on the platform on Speech Day but on a social footing. An annual event is a good start—a tea-party at the beginning of a new school-year or 'an annual post-Speech Day cocktail-party'. Such meetings help to bring about the happy state of governors 'backing new schemes and providing financial backing where possible'. Better that than the situation where 'the majority of the staff do not know even the name of any governor'.

The sheer size of a comprehensive school can make life different for a man's first year in such a school. Teachers outside such schools can develop a phobia, a fear of large numbers. Teachers inside accept the necessity, balancing advantages against disadvantages. One man includes among the advantages 'greater social possibilities. Far greater scope for experiment, for personal initiative, and for offering personal interests to children through clubs.' It is admitted that a new member of the staff is likely to feel bewildered at first.

Staff find initially that they have the same difficulty as pupils in growing accustomed to the size of the school and tend to feel rather lost at the start. They, too, rapidly settle down in most cases, although some continue to find the large numbers of children and the consequent bustle too great a strain, especially if they have come from the quieter atmosphere of a small grammar school.

One member says, 'It is not possible for staff to know all their colleagues reasonably well'. Another admits: 'After a year I certainly do not know all the staff. A staff-room of 65 is the one place in the school where I am aware of the size.'

Unity and variety

The leaders in a school know full well the danger that a large staff may split up into small groups; they take steps to counteract it. At one conference we heard that of all the ways that had been tried at one school, nothing compared with a staff concert. In other places staff societies for special interests have been tried, as well as recreational mixing—as with a club for badminton or tennis.

The tendency to split seems inherent:

With so large a staff (over 100) there is a tendency for groups to form, more especially departmental groups with similar interests. As the staff is so large, it is rarely possible for all to assemble in one room. Nevertheless, a very good spirit exists and there is much interchange of ideas and experiences between the individual groups, which themselves tend to change and to reconstitute themselves from time to time.

The layout of buildings can be an obstacle to a sense of unity in the staff. This is not confined to the most unfortunate cases where different parts of the one school are half a mile apart or more. If the Lower School is in its own building, at morning break the staff in that building will stay there and will not join the staff of the Upper School, in a separate building. The House-based school encourages the staff of one House to teach in one building and to spend their breaks in that building. It requires an effort to break out of the House circle. For the House staff to dine with the pupils in a House is admirable but it prevents the House staff from meeting the staff of other Houses over dinner. Many schools praise the provision of a staff dining-room partly because it brings the staff together and enables much to be discussed informally. That is better than calling special meetings or increasing the number of information-loaded pieces of paper for the staff.

The provision of staff-rooms has its effect on the life of the staff. One room big enough for all the staff to come together in some comfort at break and at dinner-time—that is something that many schools lack. One staff-room, even one big enough, is not sufficient: a mixed school needs a room for mistresses only; a school with only one staff-room has not provided a quiet working place for free periods. Three staff-rooms offer variety but may be 'a social inconvenience'. Three staff-rooms may demand three visits by the Head of Department who needs to see those in his team on an urgent matter.

Several ways have been tried of bringing the staff together, apart from a staff dining-room and social activities already referred to. One school has made it a rule to meet every Friday for tea and a short meeting. Other schools fix one day a week for a dinner-time coming-together or for a more formal meeting—'The staff hold a meeting at lunch time every Monday, at which any teacher can express his point of view'. More will be said later about the function of a Common Room Committee or a Staff Council.

The size of the staff could be a temptation to passengers.

On junior members of staff, the insidious effect which could be dangerous to morale is that of feeling like a very small fish in a very large pond—and that one's efforts (or one's shortcomings) might go unnoticed. Particularly in the matter of staff absences, whereas in a smaller school a teacher might be reluctant to be absent because of the heavy load of work thrown on to colleagues in consequence, he or she might be less reluctant in a large school —in the knowledge that lesson replacements are such an impersonal matter, and in any case there is normally a fair number of 'spare' staff to fill in the necessary periods.

This correspondent is not alone in thinking that size makes it more difficult to achieve and maintain high standards of discipline. 'Individual staff, sometimes far from the centre of authority, escape the censure that a smaller, more tightly-knit staff might impose.' Bigger variations in personal standards will include bad examples for younger members to see and to be tempted to follow. There can be, too, variations in degree of readiness on the part of those in authority to tackle colleagues who turn a blind eye or keep away from sources of trouble. The experience of a third correspondent is: 'When the whole school moves between lessons, one feels somewhat swamped. I have noticed (in two different schools) a tendency for staff to abandon the task of corridor supervision and to retire to their own classrooms where control is easier.' It comes to this—the bigger the staff, the more scope for the human factor, the bigger the test of character of the individual teacher.

A teacher moving from a grammar school to a comprehensive school must be prepared for a bigger range in the bigger staff. We do not approach this in terms of graduate and non-graduate. In a comprehensive school a teacher is valued according to his performance, not according to his form of training. A bigger staff means more teachers on probation, more teachers who are developing and consolidating after the year of probation. There will be more young teachers—a good thing. There will be no one day in the week when all the staff are present, because of the bigger proportion of part-time teachers. There will be more people with experience of the way things are done in other schools, partly because of the number of part-time teachers. Practical subjects will play a larger part than in a grammar school; put together all who teach Art, Music, Woodwork, Metalwork, Technical Drawing, Typing, Domestic Science, Physical Education and you have a substantial core of teachers. These, moreover, do not move to find their pupils. Their pupils come to them in their specially-equipped rooms; they remain there for most of the day except for dinner and are less likely to be found in common rooms during short breaks. If you need to see them during morning interval you track them down where they work—and their work-places can be in the remotest parts of scattered buildings. You will meet 'educational adventurers and idealists', those who are devoted to the comprehensive concept, those who want to stay in the school and work out their ideas and notions, and those who put up with what they do not really like until something better turns up. Variety of staff can give spice to life in a comprehensive school.

The Headmaster

It is completely unrealistic to think of grammar schools of 1966 as units where every Headmaster knows every child. One-third of the comprehensive schools about which we have detailed information are no bigger than top-size grammar schools. There must be a point—it may be 500 or 600—beyond which a Headmaster could not know every pupil in his school unless he gave all his time to getting to know them. Let it be admitted that the Headmaster of a comprehensive school at full strength cannot know his pupils as well as the man in charge of a school of under 500 pupils. Add that the personal influence of the Headmaster cannot be felt in a comprehensive school to the same extent or in the same way as in a school of under 500 pupils. This applies equally to the Headmaster of a grammar school of more than 600 pupils. Some people worry about how accessible the Headmaster is to his staff. Again, it is the same for all schools with 600–1,000 pupils, grammar or comprehensive and it is a bigger problem with schools of over 1,000 pupils. 'The Head is always only too willing to see his staff, providing he is in the school and can be found' could be said of many schools of different kinds.

The teachers in comprehensive schools appreciate the difficulties and admire efforts by Headmasters to deal with situations arising from size.

The personal influence of the Head is enormous, because he makes sure he is accessible at all times, stays in the Staff Rooms every day for long periods just to chat to staff, and teaches all first-year children. His influence may not work in the same way as Heads of small schools, but because day-to-day administration is all delegated, the Head devotes his energies to 'Public Relations' with staff, parents, and visitors, and to policy-making. Most of the new ideas started in the school were started by the Head during the past five years.

That is representative of good schools. 'The Head has an open door for staff and pupils alike' can be set against 'Individuals who wish to see the Head may find themselves queueing', but the latter is an exceptional complaint. From the Headmaster's point of view there is more of everything: more staff, more specially-equipped rooms to which he must go to discuss detail on the spot, and, under a vigorous House system, more (House) dining-rooms to visit in turn, more (House) common rooms. Yet of one Headmaster the verdict was 'He is rarely at a loss over what is happening in the school, despite the apparent remoteness of his position'.

The Headmaster's influence on the staff may begin with the time-table which carries out his ideas of what is best for his children, may continue with the duplicated material drawn up to make the picture of the school programme clear to the staff, may be strong in the sectional assemblies which he takes in turn, may be most direct in staff meetings. The Headmaster must work through delegation and organization. He naturally chooses for his Deputy (and for the next in importance, Senior Master and/or Senior Mistress) someone who shares his ideas and beliefs, someone who can speak and act for him because he knows as by instinct what the Headmaster himself would say or do if the problem were put to him. The position of those next to the Deputy is much more clearly defined in a comprehensive school than in a typical large grammar school. Head of House, Head of Department—each means much more, involves the person in greater responsibility and more work than in a grammar school and thus each becomes a channel for the Headmaster's influence. With a combination of Housemasters and tutors, a thorough and effective system can operate:

House tutors are responsible to Housemasters for the day-to-day registration and collection of dinner money and dissemination of information to their tutor groups. Housemasters are responsible to the Head for the pastoral care of the children in their Houses, and are responsible for the discipline of, and advice given to these children. Heads of Departments are responsible for the organization and teaching of their subjects, and help with the advice given by Housemasters. Housemasters and Heads of Departments are therefore responsible to the Head for the routine day-to-day running of the school. Most problems needing more weight are referred to the Deputy Head for the boys and to the Senior Mistress for the girls.

Another correspondent refers to the last two office-holders as 'filters'. There is no doubt that a Headmaster can make and use ordered delegation so that his faith, his outlook, can be felt through all the activities of the school. But a large school needs personality in its leaders; a school 'easily loses coherence and confidence unless the leadership is strong'.

Consultation

From only one school is there a complaint about little consultation, 'the biggest cause of frustration' in the staff. It may be that those consulted by the Headmaster, the intermediaries, are negligent about consulting the rank and file. The general picture is one of frequent and regular consultation with 'the leaders', Heads of Houses, Heads of Departments, Heads of separate Schools. When the occasion arises, the Headmaster calls a meeting of Heads of Departments to discuss teaching matters, a meeting of the other Heads to discuss pastoral care and matters of organization apart from lessons. The way it works at one school is this:

The school is run very much as a team venture and the staff concerned are invariably consulted before any major changes are made. The more formal machinery is the House Committee (all Housemasters and Housemistresses), Academic Committee (all those in charge of subjects) and Staff meetings. All these meet at least twice every term.

At another school, the pattern is similar:

(i) The whole staff meets together about five times a year for formal discussions.

(ii) A 'Panel' of Heads of Departments and other senior staff meets once a month for about $1\frac{1}{2}$ hours after school. Any member of staff interested can in fact attend these meetings.

(iii) Housemasters meet the Head for 'working lunches' once a week (one day for Middle School and one for Upper School).

(iv) House tutors meet once a fortnight during lunch hour.

(v) Subject Department staff meet once a fortnight.

It should be apparent that a teacher cannot take an active part in policy-shaping and school government without giving up considerable free time. Partly because of the time-factor, regular meetings of the staff within a Department are the exception. 'The wide range of activities and duties involved' makes it difficult to find a time when all the members of a Department are free. In another school,

Small Departments with only a few staff meet informally and easily within their part of the building. A few Departments (Art, Technical) meet informally during morning break with their own tea/coffee arrangements. For some larger Departments meetings are difficult and possibly rarer than desirable. There is no time-table provision for subject staff to be free for a period together, and the numerous dinner-time and after-school activities make it wellnigh impossible for ten or eleven members to be free at the same time.

One of the dangers in a large school—not necessarily confined to comprehensive schools—is 'the separation of staff into Senior (decision-making) staff and the rest'. The price of unity is eternal vigilance and sacrifice of time to bring the other ranks into the picture. No sympathy is expressed for those who complain about decisions but who failed to attend voluntary meetings where they were discussed or who attended a meeting and remained silent. There is some sympathy for those teachers in a school with restrictions. At one school, Heads of Departments 'are not permitted to discuss the Headmaster's policy unrequested or to discuss matters' that are considered the special field of Heads of Houses. There is sympathy for those who find it easier to receive information and direction from the top than to get their views through the 'filters' to the man at the top. There is sympathy for the man who at the end of a long staff-meeting finds everyone else wants to go home when he wishes to bring some point up for discussion. In such cases the complaint can be expected: 'Insufficient discussion of educational topics and problems; staff meetings concerned with administrative points; too much emphasis on trivial matters'.

For some years now in some schools a Staff Council has operated successfully as a means of expressing the views of the staff. In one school this is in addition to permanent committees of Heads of Departments and of Housemasters, each with its chairman and secretary maintaining liaison with the Head, 'who would attend the meetings if necessary'. The Staff Council includes the Head, the Deputy Head, the chairman and secretary of each of the two committees, and ten elected members. Each House block (two Houses) elects one member. Five more are elected by the staff as a whole, without regard to House. 'This Council meets monthly and any members of staff may submit items for discussion and decision, and attend in person to present his case. This Council is the main policy-making body and can over-rule the Head.' At the other end of the scale is the school where 'little notice is taken by the Head of the Common Room Society (almost 100 per cent staff membership) but this organization makes itself felt by individual members of its committee raising subjects at staff meetings'. As the Headmaster's personality is so important, each crew must decide on the way of working with and under its captain.

Demands of the job

At one of our conferences, the chairman asked, 'Is there more strain in a comprehensive school?' The answer was prompt and powerful—'Very much so'. First, the teaching itself can take more out of a teacher. It is almost axiomatic that a man who applies for a post in a comprehensive school is ready to teach over the whole ability range and to teach those years for which his ability and experience have fitted him. (The specialists who look after Remedial Classes are exceptions; they confine their attention to the very backward.) 'The School, or the Headmaster, does not need to lay down any particular policy about the range of pupils taught', partly because teaching through all ranges is almost a fundamental principle. Demands of sixth-form 'A' Level work in a subject like Mathematics may limit the number of periods the Head of Department can find for teaching lower down the school but the principle of sharing the work at all levels of ability remains. You will not find a man who just moves from A form to A form to A form.

In four successive lessons a teacher may meet an unstreamed first form, a general upper sixth, a lower C.S.E. set, and an upper 'O' Level set. To prepare for and to adapt himself to such differing levels in quick succession causes some strain. More thought, patience, and ingenuity may be required to find the right course, the right approach, the right pace for classes of less able children.

One correspondent admits greater physical tiredness from teaching in a comprehensive school. The layout of a school may allow a man to do all his teaching in a few rooms close together. It is more likely to cause him to move from building to building, or along corridors and up stairs from one room to another in an elongated building. It may involve carrying about books (his own books and pupils' books).

A teacher must be ready to do his share in the battle against noise. He cannot quite escape responsibility for noise immediately outside his teaching space. If he moves from room to room, the behaviour of pupils on the move is still his concern. As it is impossible to know well more pupils than he teaches (or meets in out-of-school activities), it is likely that in any unruly crowd the majority will be strangers to him, and he will be something of a stranger to them. In addition, there will be the weeks when it is his turn to take dinner duty or corridor duty or dinner-time supervision. Though there are more to share the duties in a comprehensive school, there are more duties to share, partly because there are more buildings, corridors, pathways, stairs used by pupils. Even with efficient prefects, the duty of the teacher remains.

Teaching and keeping order are only a beginning. With pastoral care diligently delegated, each member of the teaching staff has to concern himself with the welfare and happiness of a group of pupils. A group-tutor may be much more than a marker of registers; active concern with thirty or so as individuals is required week in and week out. As Year-tutor or House-tutor a man has much to do towards records and reports of each of his little flock.

Communications are time-consuming, energy-absorbing yet indispensable. First thing in the morning, a teacher must see, digest, and maybe copy, those orders of the day that concern him as form-master or tutor or House member or teacher of a subject. Whether certain pupils get information about what to do or where to go that day depends on him. You cannot have a thousand or more pupils on the move over scores of rooms without organization and dissemination of information. However much morning Assembly is used for notice-giving, there is still much for the teacher to take in and pass on. The more completely essential information is given in advance and arrangements are made for collecting information from pupils, the freer the teacher in his classroom is from the interruptions that 'ruin the lesson' and vex the enthusiast for his subject. 'Pressures are much greater; even comparatively trivial matters have to be carefully organized.' Whatever is done in school, in communications as in everything else, is done on a larger scale in a larger school. 'It is hard work looking for someone' is one opinion from a man in a school well equipped with loudspeakers and telephones; this applies whether the one sought is a colleague or a pupil. Continuous selection, to be efficient and just, means putting on paper marks and remarks—everything that has to be taken into account when a change is discussed. With all the paperwork, the final decision may be arrived at during a meeting. That is but one cause of the many meetings that must be attended. If the teacher plays any part in clubs and societies, House contests, games and sports, he will be occupied during his dinner-time or after the end of afternoon school or on Saturday morning. There is no one yet who can report on 25 years of continuous work in a comprehensive school. The onlooker while admiring the best of the young and the maturing ones for their unspared efforts in so many directions wonders for how long they can keep up the pace. The appreciation of one's efforts by Headmaster, by colleagues, by parents, by pupils is some compensation, and so is the feeling that one is justifying one's belief in comprehensive schools.

Leavers and arrivers

It used to be held against comprehensive schools that the staff changed more frequently than in other schools. That may have been so when amalgamations to form a comprehensive school meant taking in the staff of an absorbed school along with the scholars. Without comparative figures for the leaving rate in grammar schools in July 1965, there is little point in offering figures for the rate in comprehensive schools. The causes for the departure of staff, where known, seem to have very little to do with the comprehensive system. In any case, the teaching profession was more mobile in 1965 than it was in 1959. Burnham has been an influence; once a young man looked round for a post with a special allowance above basic after a few years in his first post; now the search begins after a few terms.

The factors affecting staff leaving include:

(i) Most comprehensive schools are mixed. The women who start teaching are not infrequently already married. When the husband changes his residence as well as his job, the wife leaves. Starting a family begins at an earlier age than it used to.

(ii) There are more part-time staff in a comprehensive school than in other schools, including wives who will move when their husbands move to better jobs.

(iii) Bigger centres have fewer leaving because of the greater chance of changing the job but remaining in the same house.

(iv) The area. One school attributes its success in keeping staff 'to pleasant locality rather than the type of school'.

(v) Counter-attractions. The cities particularly offer attractive openings to graduates—change of job without the expense of moving house. We hear from several schools of staff leaving 'for more lucrative and less hectic commercial appointments'.

(vi) Dissatisfaction under Local Authorities who make life harder (through a mean staffing-ratio—fewer free periods, and through inadequacies of equipment) or who do not allow either as many special posts or as high grades in special posts as neighbouring Authorities.

(vii) Happiness. There are purpose-built, happy schools where 'only two or three leave each year and these are on promotion or retirement'.

It is more profitable to think in terms of continuity. Rather than count the birds of passage, consider how many of the key-posts are held by staff who stay. One school reports, 'Of the present full-time staff of 71, over 40 have been here for five years or more'. In another school, 25 per cent have been there for 15 years and another 25 per cent between 6 and 15 years.

No school reports difficulties of recruitment except in subjects where there is a national shortage (e.g. Mathematics) or in posts where it is reasonable to expect additional payment and none is available. An area attractive to live in has the biggest field to choose from. A school can have a reputation for being a happy school or an adventurous school—and no lack of applicants. There are some idealists and more careerists seeking experience in comprehensive schools. A study of reasons for leaving grammar schools would show how many see comprehensive schools as a field with opportunities for advancement. The current of opinion in Colleges of Education and in University Departments of Education is not against teaching in comprehensive schools. A well-planned, well-built new comprehensive school can seem 'so superior in amenity to the average grammar school'.

Additional payments

What is the attitude in comprehensive schools towards internal promotion? The answer is that there is no 'policy'. A wise Headmaster, in any type of school, selects the best of the field. If the internal applicant has the experience and the qualities, it is not held against him that he is already in the school. In growing schools, with a younger-than-average staff, the need for men of experience can often lead to the appointment of an external applicant. In one reported case of a good man being passed over, it was thought that the attitude of the Governors was the explanation.

The key-posts in organization involve great responsibility and hard work. A Housemaster in a comprehensive school has very much more to do than his counterpart in a grammar school. A Head of Department in a comprehensive school has more to organize, not merely because there may be more forms, but chiefly because the range of ability in any one year is much greater. A Head of Lower School or of Middle School or of Upper School may be in a separate building and almost the Headmaster in his school. A Deputy Headmaster is like a Managing Director in his multifarious responsibility for the smooth day-to-day running of the school. One thing which all deserve is a time-table appropriate to the load. Usually they get it. The teaching load is reduced to make possible the administrative load. In one school the Deputy Headmaster does no teaching; in no school of over 600 does he teach more than half the time. A generous staffing-ratio is essential if justice is to be done; where it is ungenerous it is the Housemaster who suffers. In terms of payment, the post in a comprehensive school which carries more work and responsibility than a similar post in a grammar school should have at least as much in additional payment. We regret that this does not always happen.

In any school it is a difficult and individious task for a Headmaster to allot the posts and payments to cover the needs of the school and the deserts of the staff. If he is a Solomon, and he thinks that one man in a key-post deserves a large Grade E allowance, he is baulked if the Local Authority refuses to allow any posts at that level. If the school is a growing school, its numbers are rising faster than the rating based on unit-total. September in each school year in a growing school makes the unit-total of April of that year, on which extra allowances are calculated, out of date and the allowances based on the smaller number inadequate for the larger number. In our figures for English schools, therefore, we separate those schools opened in 1960 or later.

In a growing school 'doubling' may be inevitable, so that where there are not enough posts for the work of Head of House to be recognized by itself, a combination of Head of House and Head of Department is made (as in School 14 of Table 7). This is usually a temporary arrangement; it is recognized that such a double responsibility in a full-grown large school is too much. So, one post may be for Remedial work and Library duties together. In the school providing these examples (1,200 pupils) other posts were allocated in July 1965 as follows:

Grade E   Head of Science, Maths, English.
    D   Head of Mechanical Science.

Grade C   Heads of Houses; Heads of History, Geography, Modern Languages, Physics, Chemistry, Commerce, Applied Maths.

     B   Heads of Boys' P.E., Girls' P.E., Art, Rural Science, Latin, Domestic Science.

     A   Music, Scripture.

Scale III   Nil

     II   Metalwork, History, Science, Maths, French, Remedial Work, Geography, Science.

     I   Cookery, Art, French (2), Science, English (2), P.E. (2), Commerce, Woodwork, Library.

Of these posts, 26 were not held by graduates or those with qualifications equivalent to a degree. In reserve were these posts, to be allocated as staff gained experience: English (2), Drama, Remedial, Maths (2), Science (2), Art, Commerce, Needlework, French.

Details of posts in a school with just over 2,000 pupils follow:

Grade E   Heads of English, Science, Maths, Languages. Sixth-Form Master.

     D   House Masters. Head of Physics.

     C   Heads of medium-size Departments (Art, Boys' Craft). Second Maths, Assistant Head of Lower School.

     B   Heads of smaller Departments (Geography, History, Commerce). Second Languages.

     A   Small departments (Religious Instruction).

Scale III.   Lower School Housemasters. Second English, Physics. Third Maths.

     II   Heads of major Departments in Lower School. Small sub-departments (Needlework).

     I   Assistant Housemasters in Upper School. Assistant Sixth-Form Master. Heads of smaller Departments in Lower School. (Typing.)

These gradings should be aligned with the details of teaching periods (a 39-period week).

13 Deputy Head, Senior Master, Head of Lower School.
20 Sixth-Form Master.
26 Housemasters. Assistant Head of Lower School.
30 Lower School Housemasters.

The general picture is one of efforts to reward those who have responsibility and make some special contribution. Apart from standard subjects, each of the following earns an additional payment in some school:

Administration
Agriculture
Backward Pupils
Boys' Brigade
Choral Music
Closed-circuit Television
Commerce and School Bursar
Co-ordinator of Sciences
Counsellor
Craft, Heavy
Craft, Light
C.S.E. Organization
Drama
Duke of Edinburgh Award
Film
Finance
Fourth-form Leavers
Further Education
Head of Newsom

Home Economics
Language Laboratory
Machine Shop
Modern Studies
New Entrants
Non-exam Fourth Form
Office Arts
Orchestra
Outdoor Activities
Rural Studies
Scouts
Sixth-form Mistress
Slow Learners
Social Sciences
Swimming
Time-table
Tuck-shop
Visual Aids
Workshops

Not only the work done, but the description of the work done too, clearly indicates the individuality possible in a comprehensive school.

Table 6. Distribution of Special Posts and Special Allowances, July 1965. English Schools opened before 1960

No.	Sept. 1965 Pupils	July group	Heads of House No.	Heads of House Grade or Scale	Head of School L. or J.	Head of School M.	Head of School U.	Sci.	Maths.	Eng.	Mod. Langs.	Hist.	Geo.	Careers	Reme-dial
1	710	XIX	—	—	C	D	—	B	C	C	—	B	B	—	B
2	710	XVIII	—	—	—	—	—	C	C	C	B	B	B	—	A
3	800	XIX	—	—	—	—	—	D	D	D	A	A	B	—	A
4	810	XIX	—	—	D	—	—	C	C	C	C	—	—	—	A
5	910	XX	—	—	—	—	—	D	D	D	D	—	—	—	I
6	950	XX	—	—	D	—	—	—	D	D	C	C	C	I	A
7	960	XXI	3	C	—	—	—	D	E	D	D	III	C	—	—
8	960	XXII	—	—	—	—	—	D	D	E	C	C	C	III	I
9	970	XXI	—	—	C	—	—	D	D	D	B	B	B	I	—
10	980	XXI	—	—	—	—	—	D	D	D	—	C	B	II	I
11	1,110	XVIII	{ 1, 5 }	{ III, II }	C	—	—	B	C	C	A	B	B	—	—
12	1,150	XIX	{ 2, 3 }	{ B, C }	—	—	—	C	C	C	B	B	B	I	A
13	1,220	XXIII	—	—	E	—	—	D	E	D	D	C	C	—	B
14	1,240	XI	{ 1, 3 }	{ E, B }	—	—	—	E	D	D	B	C	B	—	A

144

No.	Value														
15	1,250	XXIII	8	C				E	E	E	D	C	C		III
16	1,300	XXIV	14	C				D	D	D	D				II/I
17	1,310	XX	—			D	D	E	E	E	C		A		2 of II
18	1,340	XXIII	{10, 10}	{I, CJ}				D	D	E	D	B	D†		—
19	1,360	XXIV	3	C				D	D	D		B	B		—
20	1,380	XXIII	{2, 1, 1}	{B, III, II}	E			D	D	E*	B	B	B		I
21	1,420	XXIII	5	D	E			D	E	E	D	C	C		III/I
22	1,430	XXIV	—		E				E	E	C	C			A
23	1,450	XXV	—		C	E		E	E	E	C	D	D		—
24	1,580	XXV	—					E	E	E	D	C	D		—
25	1,650	XXIV	—						E	E		B	B		—
26	1,940	XXIV	{10, 10}	{C, I}				E	E	E	E	D	D		C
27	2,030	XXV	D	D				E	E	E	E	B	B		I
28	2,040	XXIV	{6, 2}	{B, CJ}	E			E	E	E		C	C	I	—

* Combined with Head of House. † Combined with Head of Sixth.

TABLE 7. *Distribution of Selected Special Posts and Special Allowances, July 1965. Growing schools*

No.	Sept. 1965 Pupils (to nearest 10)	July group No.	Heads of House No.	Heads of House Grade or Scale L or J.	Head of School L or J.	Head of School M.	Head of School U.	Sci.	Maths.	Engl.	Mod. Langs.	Hist./Geog.	Careers	Remedial
1	510	XV	—	—	—	—	—	—	B	B	B	A	—	—
2	600	XIV	—	—	—	—	—	B	B	B	—	—	II	—
3	680	VIII	—	—	—	—	—	C	C	C	B	A	—	—
4	870	XV	—	—	D	D	D	C	C	—	B	B	—	—
5	910	XXIII	1, 8	III, C	—	—	—	E	E	E	D	—	—	—
6	1,040	XVIII	—	—	D	—	D	C	D	D	C	C	—	A
7	1,160	XXII	—	—	—	D	—	D	D	D	C	C	—	A
8	1,200	XXIII	8	C	—	—	—	E	E	E	C	C	—	II
9	1,210	XXII	5	B	—	—	—	—	E	E	C	C	—	—
10	1,250	XXII	—	—	E	—	—	D	D	D	C	C	C	B
11	1,310	XXI	—	—	—	E	—	—	D	D	C	A/B	—	B
12	1,320	XXII	—	—	D	—	D	D	D	D	C	B	B	A
13	1,450	XXV	—	—	D	—	D	D	D	D	C	C	I	C
14	1,480	—	—	—	C	—	E	D	D*	D*	C	C*	—	—

Note. In accordance with the 1965 Burnham Report, the Head of Department and Graded Posts mentioned above had the following values: Grade E, £660; Grade D, £540; Grade C, £420; Grade B, £300; Grade A, £200. Scale III, £300; Scale II, £200; Scale I, £120. (These above scale payments are additional to the teacher's basic salary which is assessed in accordance with his qualifications and the length of his teaching experience.)

* Combined with Head of House.

12

EQUIPMENT

In terms of equipment, some schools are rich and happy, others are poor and depressed. The average position may be described as 'adequate without being luxurious'. At the best, the supply is very good but with teachers wanting to give their classes all the benefits of teaching-aids the result is 'they really *are* used to the full'. In addition to a Television Room and a language laboratory, one school of just under 1,400 scholars has available: four 16 mm. sound projectors, eight film-strip projectors, three epidiascopes. A rather larger school (1,650 scholars) employs: three 16 mm. sound projectors, eight film-strip projectors, two epidiascopes, six tape-recorders, six record-players, one sound-radio set with 12 extension speakers and a TV set in its own room.

Generally speaking, it has been an advantage to start a comprehensive school in the sixties, when it is so much more accepted that teachers are skilled in using and that pupils benefit from teaching-aids in their wide variety. For example, an older school (1,200) has two projectors and a tape-recorder: 'The committee will allow no more.' Are there administrators who regard a large school as some kind of Ark, with not more than two of one kind allowed? The supply of films and film-strips is so much better than it was seven years ago that now there is good material in subjects which formerly were not catered for. Moreover, the introduction of colour into films and film-strips has led to the provision of new material that can make more impact and arouse more interest than pictures in black and white. Teachers of English have now four times as many 'living voice' records of plays and poems as they had seven years ago. Recently, too, a good beginning has been made in publishing text-books which have records specially made to accompany them. Some schools have what they wish to use in spite of a niggardly Authority. It makes one both pleased and sad to read that a school has a tape-recorder and a projector 'provided by the Parent-Teacher Association' or purchased from the proceeds of a school concert.

The Science Department of a comprehensive school often has its troubles. A school of 1,650 pupils has the sum of £450 for text-books for one year. Call it 5s. per head and look through the catalogues of publishers to see if there are any science books for schools that cost no more than 5s. The bill for furnishing one student of Science with books for his 'A' Level subject—would £5 cover it? In Science particularly, the advance of knowledge makes text-books quickly out of date. Again, the constant revision of syllabuses in Science, for 'O' Level and 'A' Level, calls for changes in books. Similarly, a school which seemed to be well-equipped for a start in the 1950s cannot be expected to maintain its standards without a generous fund for additions and replacements. 'Much of the older equipment is ready to be replaced' is true of many schools. Apparatus once thought suitable for university work is now required for sixth-form teaching. New apparatus has been devised to accompany the expanding territory of scientific knowledge. Good apparatus can be very expensive. No wonder, then, that the Nuffield Curriculum Development Projects in Science are so much a part of science teaching in comprehensive schools. Participation in these projects enables schools to be up-to-date in apparatus, partly because the prestige of the venture acts as a lever to move the Local Authority to be ready to supply the money.

Whether a single Department in one school gets the money it needs to provide classes with enough books and good books depends on two controls: the Local Authority's grant to the school; the Headmaster's distribution of the monies received. One school of 1,150 pupils has to find text-books, stationery, library additions and replacements, audio-visual material, games equipment, all out of a total grant of just over £5,000—say 90s. per head per year. The larger the school, the more paper is needed for effective communication. The more a school tries to supply the parent with information about courses and choices, the more paper is needed. A school with growing numbers cannot buy as much as it used to with costs rising all round if the total sum available is not increased to take full account of these factors.

Under some Authorities the money for text-books is a separate calculation. These figures show how the attitudes of Local Authorities vary.

Pupils (to nearest 50)	Annual sum for text-books £
1,100	1,400
1,250	5,000
1,350	5,000
1,400	3,000
1,950	3,500

One other school has £3. 17s. 6d. per scholar for text-books, which is better than all but one of the schools whose figures are given above. If the most generously treated schools 'manage', how handicapped must be those who receive less! And 'manage' is the best to hope for. The broad pattern is for text-books to be improved by more diagrams, more pictures, more illustrations—all adding to the price. A comparatively 'plain' text, without such appeals to the eye, costs more now than it did a few years ago because of the increased cost of materials and the increase in wages. A Solomon is required to share the school's total among different claimants, combining justice and mercy. If the total is inadequate, a Solomon may be unable to prevent 'hard cases' such as that of a Science Department with £700 per annum and over 1,650 minds to feed, or a History Department with £150 for those out of 1,300 who take the subject, with the result that 'Advanced students are sharing text-books'.

Inadequate allowances for libraries are a sad feature of all kinds of schools, not just comprehensive schools. Library books, like text-books, are improving in appearance and increasing in cost. The National Book League has tried to raise the standard of allowances and to show how far some Authorities fall below their suggested minimum.* A comprehensive school with pupils of 11 to 19, pupils with the widest range of interests, pupils with differing levels of intelligence, has a special claim to a generous allowance for its library, indeed *libraries* in schools with completely separate Lower School or with six or more Houses in separate House Blocks. How Local Authorities recognize the needs of these schools is shown in these figures:

Pupils (to nearest 50)	Library Allowance £	Pupils (to nearest 50)	Library Allowance £
800	200	1,200	543†
950	165	1,200	800
1,000	250†	1,350	450
1,050	100	1,400	1,086
1,100	819	1,900	1,470

* In March 1965, a new memorandum on Book Expenditure in Schools, issued by the Association of Education Committees, recommended that the annual maintenance grant for school libraries should be £1 per head for pupils of 11–16, £1. 10s. per head for pupils of 16 and over.

† These schools have supplementary assistance from Municipal or County Library, the smaller school having 1,000 books on loan, the larger receiving a grant of £250.

EQUIPMENT

Music in schools has long ceased to begin and end with singing. Music for C.S.E. and for 'O' Level G.C.E. calls for 'texts' and for a library of gramophone records. A comprehensive school is certainly small if all the work in Music can be done by one teacher. Few schools are as well supplied with pianos as the one of 1,200 pupils which has seven. As soon as a school begins to form a band or an orchestra, money is needed, and not in driblets. One school of just under 1,000 pupils succeeded in forming a Brass Band, the instruments being provided by charity and self-help rather than by official support. Friends and parents made gifts; a few children were able to provide their own; concerts raised funds to make more purchases possible. Enthusiasts have performed wonders in providing instruments. In another school of 1,300 an orchestra of 40 players has been built up; the Music staff say they have had to 'beg, borrow or steal' instruments, a statement which we assume to be two-thirds correct. From their experience they say that for such an orchestra an initial grant of £1,000 is needed, and an annual grant of £200 for maintenance and replacements.

Sport brings its problems too. Probably a mixed comprehensive school needs more money than a boys' school of the same size. The cost of equipment has increased more than is realized—for example, what would you pay now for a cricket ball which cost 10s. 6d. in 1939? One school (1,200) pupils has for its winter and summer games and its athletics—just £500. With money being constantly needed for replacements, it has become increasingly difficult in any school to extend the range of games and sports. Many schools would like to offer their growing sixth forms the choice of many forms of physical training in the broadest sense. Boys who have made no headway in Soccer, for example, may find pleasure and satisfaction in a new sport.

A predominantly Soccer school that seeks to add a new activity such as Rugby Football, Hockey, Lacrosse, Basket Ball, faces a very, very large initial outlay. Some schools would go further and introduce camping, canoeing, rock-climbing; in all these activities what is cheap may be dangerous.

This short account of the equipment that teachers in comprehensive schools regard as desirable or necessary if the students in their charge are to develop their poten-tialities would be incomplete if it evaded the problem of maintenance. These teachers are asking their local authori-ties to buy expensive things. Often these items are so costly that it is really false economy to do nothing for—and pro-vide no money for—their maintenance. The experts will care for and protect the instruments and apparatus they receive. The stage is soon reached where it is plain that they are mere amateurs and that their material needs profes-sional attention. However fine a pianist the music master is, we do not expect him to tune the school piano. Our mech-anized schools of the future will need several competent professionals to see to language laboratories, audio-visual material, musical instruments and scientific apparatus. In the past it has been too easy to leave repairs and over-hauls to some handyman in the ancillary staff—who may be kept so busy that he has little time to do the work he is paid to do. The goodwill and skill of laboratory stewards or others should not be presumed on. The choice seems to fall between a full-time worker with his own workshop/store-room and provision for regular visits by outside workers to give professional attention. The Inner London Education Authority takes the second choice. It employs its own staff and sends into the schools as required men who attend to large apparatus in the Science Department, workshop equipment, musical instruments. It is accepted that a large

EQUIPMENT

Authority can have a team to cut the grass and can send the men and their equipment from school to school; it will soon be accepted that there is in comprehensive schools much costly material that can best be 'serviced' by peripatetic professional experts.

13

EXPERIMENTS

Some comprehensive schools must be particularly stimulating places in which to work. Apart from belief and
enthusiasm for the comprehensive principle, a permanent
feature in the oldest schools and an invigorating one in
some of the newest ones, there is often a delight in taking
part in educational experiments in selected subjects or
selected fields of activity. If there comes to be a tradition
in comprehensive schools, it is likely to be the tradition of
being ready to make new approaches and to try new
methods. Many leaders are far from content with long-
established ways of teaching the normal subjects of a
programme of book-learning. If the patterns of teaching
children between 11 and 18 are markedly different in
1976, much of the credit will properly belong to the forward-looking, unhidebound teachers in comprehensive
schools in 1966. Though the pupils in these schools are not
guinea-pigs, it must be acknowledged that all educational
experiments have to be tried out on children. Comprehensive schools are providing the children.

A school of nearly a thousand pupils can undertake a
number of experiments in different subjects. One school of
this size is working on the following:

(A) Programmed Learning.

15 Grundytutors and 18 branching programmes being
used on a phased investigation of:

   (i) the organizational impact of hardware on a
   school;

   (ii) the desirability of pairing pupils when using
   machines;

   (iii) finding the useful age/ability range for certain programmes;

   (iv) the use of programmed texts and linear programmes in conjunction with the branching programmes.

This experiment began with one third-year class/set working on 'Indices' and 'Longitude and latitude' programmes.

(B) Physics Department. Experimental 'O' Level syllabus work with Pilot Scheme of the Association of Science Education. (Experimental approach and inclusion of atomic, nuclear and other modern Physics.)

(C) Maths Department. Second year of Midland Maths Experiment. Three calculating machines used in first year and with sixth-year Statistics.

(D) Latin Department. Linear programmed text in preparation on syntax (Group of Midlands Classical Teachers).

(E) Geography Department. Research into the assessment of field work ('A' Level, J.M.B.).

(F) Programmed text-books. Crash courses for 6b (C.S.E. successes) 'O' Level candidates in Maths and Geography on topics insufficiently covered by C.S.E. syllabus.

(G) Modern Languages, including Russian. Through audio-visual courses.

One school could hardly be doing more in its determination to fit courses to pupils and to avoid offering middle and lower streams 'watered-down academic courses'.

A smaller school (about 800) apart from working TV and radio broadcasts into the teaching programme, and making a fresh approach to Cultural Studies in the sixth form, has all the first four years involved in the Nuffield Physics teaching project and will be offering 'O' Level Nuffield Physics in 1967. In its workshops, the importance of design is stressed more than in most schools. The workshops include a department for casting and forging in aluminium and iron. This school has taken part in a project, Assessment of Engineering Studies, from the start. The Art Department includes a printing section which produces the school magazine.

If in some grammar schools the C.S.E. examination is regarded as a sideline, in comprehensive schools it is more like an equal partner with G.C.E. Teachers are enthusiastic about it and relish the opportunity to decide what is to be taught and how learners are to be tested. In a year or two they will have valuable experience in operating Mode 3 in the widest possible range of subjects.

In Modern Languages, schools are making the fullest use of new methods and are active in experiments that keep them in touch with other schools. One school arranges audiovisual teaching for all its forms, not as a last resort for those who are making little headway. Behind the term 'audiovisual' there is deep and thorough study. In one school 'audio-visual' means 'courses based on word-frequency counts; study of structural linguistics; linear programmes; phased progress to synchronized cartoon images with taperecordings'. Some schools work to the B.B.C. audio-visual course; other schools take part in the Nuffield Foundation course in languages. At least one school is the local centre for experimental work in languages, joining with Junior Schools in the neighbourhood to make a start with learning a foreign language before pupils reach the age of 11. Russian, particularly, is an open field for experiment. One school wishing to apply audio-visual techniques to Russian came up against a failure of publishers to offer a suitable text-book. The result is 'the text-book is being written here'.

School after school reports taking part in Nuffield Foundation schemes for Science. Some specify the Nuffield Physics project; others have begun work with the Nuffield Biology project.

Mathematics, too, is a subject for fresh approaches. The use of an overhead projector is being tried. One school has submitted its own Mathematics Syllabus for 'O' Level and is being examined on the accepted syllabus by the Associated Examinations Board. The Midland Mathematics Experiment has several comprehensive schools in its list of members. One school is working out a new syllabus in Mathematics in conjunction with the local University.

Team-teaching is being tested in a few schools. Whereas such experiments, like those in programmed learning, seem to put first the technique of teaching, other experiments are being conducted which are more concerned with the technique of learning. Work is being done, in conjunction with a University Department of Education, on Factors of Success.

Almost every comprehensive school receives students for school practice from College of Education and from University. One venture is of particular interest:

It is envisaged that a group of students all taking the same subject (initially English) will come into the school and will be under the close supervision of the Head of Department. The latter will hold seminars with the team of students and take on much of the job usually done by the supervisor.

Such an arrangement, based on sound principles, will provide an excellent test of the value of team-teaching. In any case, such liaison between practising teacher and lecturer is something long waited for. We are glad that it begins on the right lines by the Head of Department going to the Department of Education to meet and prepare the students before they first come into the school.

The good work of the County Colleges of Cambridgeshire is widely acknowledged. One comprehensive school is doing similar work, going far beyond the elementary stage, where a Centre for Further Education occupies part of the buildings of a comprehensive school. One new school is truly the educational centre for a district. It has been planned 'to provide for the total educational needs of the area, containing as it does a Residential Centre for courses on the site under a full-time warden, and envisaging a closed-circuit TV link with the local College of Education'. In addition, part of it is a residential home for deprived children, for children from homes where conditions greatly handicap a child.

It is surely a healthy sign that so many teachers are ready to try to enlist the full co-operation of the pupil by offering something that will appeal to the young learner of today. Only an unusual child will refuse to respond to what he feels is something special, something devised for his benefit, something which may arouse envy in those who do not have the chance to participate. So at the end of the summer term one school drops the stated time-table, offers a project week with novelties and the widest range of different activities. To prepare and plan such a week calls for both imagination and industry. In another school, an effort is made to lift out of a rut those whose ability is under-employed. A junior boy who has been content to freewheel may well rise to the bait of a change, which, he knows, makes him a little different from all the others, even without a tempting title such as 'Adventure Course'. Again, planning and carrying such a course demands thorough preliminary work and more energy to keep up the enthusiasm aroused.

This is not the place to deal in detail with some of the major experiments that are to be found today in our comprehensive schools. The rival merits of streaming and no streaming are being tested thoroughly in these schools. Secondly, the challenge issued by the Newsom Report has been accepted. 'The other half' are receiving the fullest attention by people who are expressing their care in schemes that show deep thought, real understanding of children and great energy. It is always so much easier to carry on with the mixture as before than to give oneself to the untried and the difficult.

14

SCHOOLS IN WALES

The mountains and rivers of Wales affect the distribution of population which in turn gives comprehensive schools in most of the country certain features which justify devoting a separate section to them. Well-populated coastal strips can have and do have good grammar schools side by side with good modern schools. A few places in the north—such as Holyhead—can support a comprehensive school with more than a thousand pupils. With such numbers such schools are little different from English schools of the same size. Apart from two features, the problems and the successes are the same.

For the non-industrial areas of Wales there have always been the problems of distances and of the size of schools. The Welsh Intermediate School Act of 1894 illustrates the efforts made to bring what we now call secondary education to towns too small to support a successful grammar school. Indeed, rather small towns which did have a grammar school, towns without growing industries, have seen little if any rise in numbers through times which have made increase in numbers an essential for survival. In the fifties, several such grammar schools had become bilateral; in the sixties the change has been from bilateral to comprehensive. The small modern school which is merged with a small grammar school makes a small comprehensive school.

Very often there is no alternative. In many of the small schools there is in the staff a strong wish for more pupils so that more could be done. One school already takes all the children within an eight-mile radius. Increasing the circle would probably bring in very few more pupils. Rivers and mountains, and county boundaries, limit many schools to a thinly-populated area. Schools must accept the conditions and make the best of things. So they do.

Hardly any comprehensive school has buildings planned for the smooth working of such a school. Usually the buildings of the grammar school are the core—containing the office and the Headmaster's Room, from which two places come directions and direction. A description of one fits many—'a patchwork of old and new'. A main road may divide the grammar school half from the modern school half, a reminder of the separation of the past and a frustration in the present. Sometimes additions have been made to buildings but what is left of the original playing-fields has to serve the needs of a larger number of pupils. A specialist room cannot be nothing but a specialist room from Monday morning to Friday night; for example, a Craft Room (with its kiln) at one school has to act as a Geography Room when there is no class for craft.

Some detailed consideration follows of eight schools, the smallest just over 325, the largest just over 825. The average would be 550. There is no comparable group of eight small schools in the English schools which have supplied the information we asked.

TABLE 8. *Eight small Welsh schools. September 1965*

Total number of pupils	4,409
Pupils in Fourth Year	842
Pupils in Fifth Year	586
Pupils with 5 'O' Level Passes (July 1965)	193
Pupils in Sixth Form—1st year	256
Pupils in Sixth Form—2nd and 3rd years	231
Pupils with 3 'A' Level Passes (July 1965)	52
Pupils with 2 'A' Level Passes (July 1965)	50
Sixth Form Pupils (September 1965) as proportion of September 1965 total	11%

TABLE 8 (*cont.*)

Pupils proceeding to University (October 1965)	55
Pupils proceeding to Colleges of Education	74
Ratio of total number of pupils to September 1965 intake	5·5
Total number of staff (excluding Headmaster and including fractions represented by part-time teachers)	247
Total number of pupils	4,409
Staffing Ratio	1 to 17·9

The staff

Two features of the staffing are indicated by figures. First, the staffing ratio is more generous than in England. Secondly, the proportion of graduates in Welsh comprehensive schools is higher than it is in England. Of the 247 teachers, 168 are graduates.

At the end of the summer term a smaller proportion of the staff leaves Welsh schools than is to be found in English schools. Moving school in Wales almost always means leaving the district. Just as the grammar school buildings provide the core of 'the new school', so 'the new school' depends greatly on the grammar school staff. Let it be admitted that most of our members are 'grammar school staff' and that their views must be judged accordingly. Schools can be gainers when the key-posts are held by those brought up in the grammar school tradition, those who wish to uphold grammar school standards and who are ready to train new entrants to maintain traditions and standards. Everywhere there is evidence of determination to 'bring on' the most able pupils. The figures given in Table 8 show how successful these schools are in sending forward good students to Universities and Colleges of Education. Equally clear is readiness to encourage pupils to be active members of a House, members of clubs and societies. One quite small school reports 15 different societies, and nine out of every ten pupils a member of at least one.

A small unit-total rules out any very high payments through special allowances. Only a school with over 750 pupils can offer a D post. The eight schools between them had 11 of the C posts, 31 each of B posts and A posts. A Deputy Headmaster is also a Head of a Department. The Headmaster, of course, has less to delegate. He is usually the man who acts as Careers Master.

Welsh-speaking teachers are needed for Welsh-speaking children. One school reports, 'Insistence on ability to teach the subject in Welsh debars many first-class applicants'. At some schools the children can choose between a class in which the subject is taught through English and a class in which the subject is taught through Welsh. Not only are the teachers Welsh but a greater proportion of them are those who choose to teach in the corner of Wales in which they were brought up. That goes a long way towards explaining the stability of the staff—fewer inducements to leave, stronger ties with the immediate locality. These comprehensive schools contain a high proportion of teachers who know the parents of their pupils, know the pupils' background and the home conditions. In their turn, the teachers will be better known by parents as well as by pupils.

The Welsh language

English readers need to be reminded that in much of Wales the Welsh language is what the English language is in England—the native tongue, the language used in the playground and sometimes the language used in class-teaching of subjects other than Welsh. But first, just as English is in many ways the first subject to be placed in the time-table, so in Wales Welsh is a subject—a compulsory subject—for which room has to be found in the time-table. In short, the curriculum in Wales carries an additional subject; the range of subjects for external examinations includes one more subject than the range in English comprehensive schools. So, one of the key-posts in a Welsh comprehensive school is the Head of the Department of Welsh.

Provision is made for the immigrants—the few whose native tongue is English. 'English-speaking children learn Welsh as a foreign language.' Therefore in the first year, when everyone has lessons in Welsh, there has to be setting, for you cannot operate mixed-ability groups with such divergence in proficiency in the Welsh language. Therefore before time is allocated to other subjects, so many periods must be earmarked for two languages, Welsh and English. With this pressure on time-table space, a slightly narrower range of subjects is open to the individual child.

Usually it is the second modern language which fails to find a foothold. As one school reports, 'Welsh is a compulsory subject up to and including the sixth form. The only other modern language is French. The Governors recently rejected the introduction of a second language (German) as it could only be introduced if pupils could be relieved of compulsory Welsh.' The Welsh boy with the gift of tongues finds it harder to get to a University to read Modern Languages.

The need to find time for Welsh combines with the limitations of small size to make it difficult to find a place for Latin. One school has ceased to offer Latin and arranges to transfer its Latin-needing pupils to a neighbouring school (and is fortunate to have one near enough that retains Latin). There is seldom enough Latin to occupy one teacher, so that graduates in Arts may have to teach Latin as a second subject.

Some schools seem to retain basic Welsh and make Welsh

as a main subject an option to other main subjects. One school offers the choice in the third year of Welsh or Latin or Art. In another school the third year option is Welsh or French; in a third school Welsh or French or Domestic Science. This last triple option operates in a different school in the fourth year. In the fourth year of another school the choice is Welsh or Needlework or Physics.

Welsh is, naturally, an 'A' Level subject and affects the choices there. One school makes Welsh optional with French and Geography. One school reports the special position of subjects which are taught in Welsh, particularly Scripture, History, Geography.

These classes are now sitting 'A' Level. The lower 'Welsh' forms are truly comprehensive in range and ability. They enter the 'Welsh' form at parents' choice. Not all pupils who have come from Welsh primary schools opt to be taught in these subjects through the medium of Welsh. Therefore we lose from these 'Welsh' streams the very pupils for whom the scheme was devised. Scripture has a distinct advantage over the other two subjects since the pupils are acquainted with their Welsh Bibles which they have already studied in their Welsh Sunday Schools.

In a different school, the sixth-form student has to choose between French and Scripture, and if he chooses French he can start Latin.

Provision for Welsh has the effect of compelling choice at an earlier stage than is normal in England. It could be regarded as a concealed form of early specialization for by restricting the subjects to be taken at 'O' Level, it affects choice in the sixth form. In these schools examples of the following options are found:

Third Year: Chemistry or Geography
            Physics or Scripture
            Chemistry or Art

Fourth year: Chemistry or Art
Geography or Music
Chemistry or History
Latin or Biology
Book-keeping or Chemistry or Geography or Art.

One school shows ingenuity in dealing with fourth-year options.

A	B
Physics or Latin	History or Biology
History or Biology	Chemistry or Geography

This makes it possible for one pupil to have more choice than appears, for pupils of *A* can opt into the subjects of *B* which are taken at the same time, so that he can take History under *A* and Biology under *B*.

The sixth form

Each school has its own policy about the sixth form. In one school, a student must have four passes at 'O' Level before being allowed to start on a course for 'A' Levels. One school plans a Commercial Course and a Pre-nursing Course for the lower sixth. It is often found difficult to arrange other courses apart from 'A' Level work as part of general education. The main cause is that a small staff is hard pressed. Sometimes the result is that those pupils who are not working for three subjects at 'A' Level are given more free periods than they have the ambition and character to make good use of. Some of the staff would be happier if, as in the past, the sixth form was reserved for scholars. Their views are represented by one man who said, 'With dilution, the tone of the sixth goes down'.

One distinctive feature of the Welsh schools is that the different years of the sixth form have to be taught together, in many schools for most subjects, in some schools for all. In one school, one subject will have half the eight periods allotted to joint lessons, and for the other four lessons the first year is taught separately from the second. Those pupils who stay on for a third year have to join the second-year sixth for lessons. This arrangement makes life harder for both teacher and pupil and explains the frequent wish for more pupils who would enable the school to have more teachers.

Differences in pupils

Differences in pupils cannot be ignored. It is not unfair to ask what these small schools are doing for five differing groups.

(1) The Remedial group.

(2) The pupils above the former group but still equivalent to the lower streams of a medium-sized modern school.

(3) The pupils who would be the middle stream in such a modern school.

(4) The pupils who would be the top stream in such a modern school.

(5) The pupils who under the 11 + basis would be selected for a grammar school.

Regret is expressed that little can be done for the most backward children. Numbers are often too small to allow a separate small class. A specially-trained teacher of retarded children is a rarity.

In a three-form entry school the C stream is the problem, ranging from the E.S.N. child to the average child who would not quite get into the top form in a modern school. The industrious ones will climb into a higher stream and stay to take some 'O' Level subjects; most of those who have little ambition and little hope in G.C.E. will leave as soon as possible. Though more practical work, within the limits of accommodation, is provided, it is seldom possible to attempt anything of a Newsom course adapted to needs.

In Wales there is less of 'teaching through', of the graduates teaching all levels of ability. There is more of a division between those who can and do take the 'O' Level forms and those who cannot and do not wish to. 'It is difficult to arrange a coherent syllabus in an academic subject throughout the school because of differing abilities of the staff.' From another source we learn, 'The less able pupil is somewhat neglected as to a large degree such pupils are provided with a "watered-down" grammar school syllabus'.

The C.S.E. examination has not helped much yet. Several schools have made no use of it at all; a few have had a fringe entry. In History there is the grave handicap of different syllabuses. It seems that parents and pupils are slow to accept it, wanting G.C.E. or nothing. We are told that it is more difficult to find the right kind of teacher for the C stream. In one school the difficulties in separating for teaching purposes those who should take 'O' Level from those who should take C.S.E. were so great that twelve of the latter had to be entered for 'O' Level, all failing. Many of the teachers regret the 'tendency to give diluted grammar school education to the C stream'. Just a few schools make use of examinations conducted by the Royal Society of Arts and Pitmans.

The boy who would have been in the middle stream of a modern school fares better. It must be remembered that in Wales it is the tradition to give grammar school education to 30 per cent or more of the intake. One five-form entry school retains its original two forms of grammar school pupils and by the fifth year has three examination forms. During the first two or three years there is regular movement upwards into the examination forms. One man thinks his school lost something through the abolition of the 11-plus, with its 'necessary preparation in primary schools'. Ending the 11-plus, in his view, 'has considerably lowered standard on entry. Pupils of B ability particularly suffer from absence of the former "push", which is what the B child needs.' In the small comprehensive school he gets his push. More pupils are encouraged to take and succeed in taking 'O' Level subjects than would have been the case under the older arrangement of separate grammar school and modern school. Setting is advocated and used to make sure of getting the brainy pupils into the top teaching-groups. So the boy who would have been at the top in a modern school benefits from being in a comprehensive school in that he can be really extended through meeting competition and he may have teachers more experienced in working for external examinations. These schools are quite proud of the number of pupils who would not have been selected for grammar schools but who took 'A' Level subjects with success.

The selected pupil will be little affected by recent changes in organization. He goes to what was the grammar school, is taught by the grammar school staff in the grammar school tradition and maintains the school's record of sound results in external examinations. A few teachers are worried about the range of ability in the top examination form. 'There is a considerable range of ability within the one teaching-group, especially in the A stream where the potential Open Scholar finds himself in the same class as the pupil who can only just pass in one or two subjects at "O" Level. The able pupil does not lower his standards and the less able faces a challenge.' That can be balanced against 'The more intelligent pupils are handicapped by the slow progress of their less intelligent classmates'. Another form of doubt is in 'The real casualty in this system of a wide choice of subjects is the old ideal of a general liberal education up to 16'. This writer deplores the concentration from the age of 14 on Arts *or* Science subjects.

The last words on the small comprehensive schools of Wales should come from our members.

(a) 'About 65 per cent receive a grammar school education and benefit from it.'

(b) 'Our system is too academic and inflexible.'

(c) 'Everybody who has the ability is given the opportunity.'

(d) 'The comprehensive school has developed naturally in this area; it is the ideal solution in a rural area. It is quite certain that many children in this area would never have attempted "O" and "A" Levels and many pupils would have been lost to universities and training colleges if the comprehensive schools had not been established. It has abolished social distinction and helped to create a more unified community.'

GLOSSARY

This is included for the information of readers abroad who may be unfamiliar with some of the terms used.

(1) *Examinations*

The 'Eleven Plus' is a selection examination taken by most pupils at the age of 11. It provides a measure of a child's intelligence and attainment and thus determines whether he receives a grammar school education or not.

The G.C.E. 'O' Level examination is taken at the age of 15 or 16 by grammar school pupils and the more able pupils in other secondary schools. The G.C.E. 'A' Level examination is taken some two years later at the end of the normal sixth-form course, mainly by prospective university entrants.

The C.S.E. examination is intended for those pupils of average and above average ability who may not be able to cope satisfactorily with the G.C.E. 'O' Level. It was intended to replace, for pupils of school age, examinations of a similar standard such as those of the Royal Society of Arts, the Union of Educational Institutes and the College of Preceptors.

(2) *Schools*

A 'Modern School' is one which normally provides education for those pupils who are not in the top 20 to 25 per cent of the ability range.

A 'Direct Grant School' is normally a grammar school governed not by a Local Education Authority but by an independent Board of Governors. Some financial assistance is received direct from the Government. There are some fee-paying pupils and there is very strong competition for the free places in these schools.

'County Colleges' have been envisaged as a possible future development providing for compulsory part-time education for those pupils who have left school at the minimum school-leaving age.

(3) *Other phrases*

'Newsom children' are children of average and below average ability. The Newsom Report suggested improvements in the education of this group.

'The Leicestershire plan' divided secondary education in the county into two stages. The 'Junior High School' takes all pupils from the age of 11. Pupils are able to transfer at 14 to the 'Senior High School' providing that parents agree that they should remain at school for at least a further two years. The rest of the pupils remain at the 'Junior High School' until they leave school.

'The Plowden Committee' was set up to consider and report on primary education. It is expected that one recommendation will be concerned with the best age of transfer from primary to secondary education.

In 'Nuffield projects' changes in the Science and Mathematics curriculum and examinations are being tried out. These experiments have been fostered by grants from the Nuffield Foundation.

INDEX

INDEX